The Care and Feeding of Transmission Lines

Joel R. Hallas, W1ZR

A radio amateur's guide to understanding transmission lines

Cover Design:
Sue Fagan, KB1OKW

Production:
Michelle Bloom, WB1ENT
Jodi Morin, KA1JPA
David Pingree, N1NAS
Carol Michaud, KB1QAW

Copyright © 2012 by

The American Radio Relay League, Inc.

Copyright secured under the Pan-American Convention

All rights reserved. No part of this work may be reproduced in any form except by written permission of the publisher. All rights of translation are reserved.

Printed in USA

Quedan reservados todos los derechos

ISBN: 978-0-87259-478-4

First Edition
First Printing

Contents

Foreword

Chapter 1 — What is a Transmission Line and Why Do We Need One?

Chapter 2 — What Are the Types of Transmission Lines?

Chapter 3 — Let's Examine Coaxial Transmission Line

Chapter 4 — Other Types of Unbalanced Transmission Line

Chapter 5 — Let's Examine Balanced Transmission Line

Chapter 6 — Transmission Line Connectors

Chapter 7 — Installing Coaxial Connectors on Cable

Chapter 8 — Determining Which Line is Best Suited for a Particular Application

Chapter 9 — Application and Installation Notes

Index

Foreword

While radio equipment and antennas get most of the amateur's attention, arguably neither can amount to much unless interconnected by a transmission line. Unless the appropriate line type is selected and properly installed, the amateur is likely to be dissatisfied with the results no matter how heavily invested he or she is in what happens at each end of the line.

This book is intended to introduce readers with a basic understanding of radio equipment and antennas to the ins and outs of transmission lines. Subjects covered include an introduction to the various types of line available, detailed discussions of the key parameters of both coaxial cable and balanced line types as well as the types of connectors and how to install them. With that knowledge, we then go into how to select the best line for each application, then how to install the line in a real environment and make sure it will still work at peak efficiency after it has been in place for some time.

As with all ARRL books, you can check for updates and errata, if any, at **www.arrl.org/notes/**.

Chapter 1: What is a Transmission Line and Why Do We Need One?

A transmission line is a kind of cable designed to convey power between physical locations. This kind of system is used for transmission of water, gas, sewage, electrical power, telephony signals and, in our case, radio frequency (RF) energy.

As with each of these applications, the desired results are the same: To convey the material or signal from one location to the other with minimal loss and with an output that includes all the characteristics of the input. While the ideal line is easy to imagine, real systems do not ever quite measure up.

In a pipe-based system, for example, material is lost through leaks in the pipe wall or at connectors. Friction of the fluid against the pipe wall results in a reduction of velocity and pressure over distance. Excessive pressure can result in a catastrophic rupture and loss of system functionality.

In a radio frequency transmission line, we have the same type of potential departures from an ideal transmission system, and some other issues that plumbers don't routinely worry about. The system functionality, however, is still very similar.

WHY DO WE NEED TRANSMISSION LINES?

We generally need a transmission line to carry RF between different parts of a radio system. **Figure 1.1** shows locations within a typical Amateur Radio station in which transmission lines would be appropriate. While not every amateur station includes all of these blocks, and some may have multiples of some blocks, the sketch is typical. Some larger pieces of equipment also have transmission lines within the unit for subassembly interconnections.

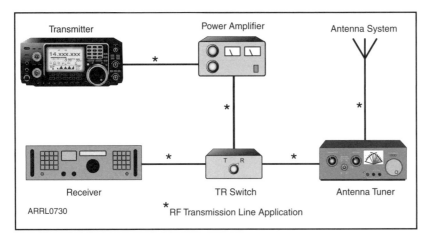

Figure 1.1 –
Block diagram of a typical Amateur Radio station, showing places that would be appropriate applications for an RF transmission line.

While we may associate transmission lines with antenna connections, Figure 1.1 indicates that they have wider application. The short interequipment lines are sometimes referred to as *jumper cables*. Note that the dc power interconnections not shown between each piece of equipment and its power source could also be considered a kind of transmission line. Power transmission is a different topic beyond the scope of this book.

The reason transmission lines are used for these applications is that if other types of connection arrangement are used, we are likely to have radiation from the connecting wires. While the radiation does reduce the signal available at the end of the interconnection, it is usually more significant that the radiation will couple to other circuitry. This can result in a number of problems, including oscillation, transmitter lockup, and interference to reception of other systems such as computers.

BUT SOMETIMES WE DON'T NEED TRANSMISSION LINES

Actually, we don't always need a transmission line for a radio system. There are some cases in which all parts of a system are in a single location. In that case, no transmission lines are required. One example of such a system that I once worked with is shown in **Figure 1.2**. The problem was to accurately determine the reception pattern of a complex HF antenna array used as a part of a military over-the-horizon radar system. The solution was to monitor the receiver output as a helicopter rode in a circle around the array with a hanging center-fed dipole suspended below. A battery-powered low power transmitter was located at the center of the antenna and connected directly to each side of the antenna. A technician at the array center took azimuth data so we could tell the azimuth of the helicopter, while an onboard

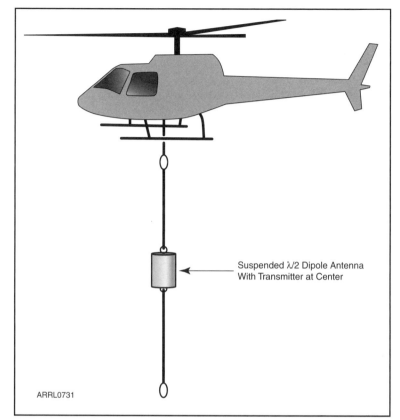

Figure 1.2 – No RF transmission line required. This small battery-powered transmitter is located at the center of a directly fed dipole. The system was used to test the receive antenna pattern of a vertically polarized directive HF antenna array used in a military over-the-horizon radar.

Suspended λ/2 Dipole Antenna With Transmitter at Center

navigation system was used to maintain a constant distance — worked like a charm!

Another more common application is found in some microwave systems. As we will discuss later, losses in transmission lines go up rapidly with frequency. Radio systems at microwave frequencies are usually designed to minimize transmission line length. Some have the actual microwave portion of the radio system located within the antenna structure. This is common in satellite TV receiver systems, as well as in microwave communication and radar systems. It is often more economical to have the interunit cabling carry the dc and intermediate frequency signals (on transmission lines, but with less loss because the signals are at lower frequencies) than to suffer the transmission line losses at microwave frequencies. **Figure 1.3** illustrates a typical configuration.

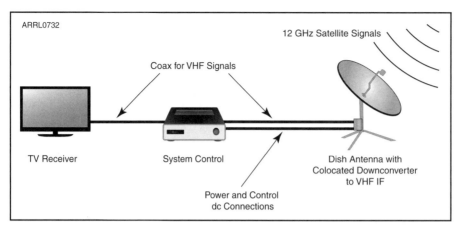

Figure 1.3 – In this simplified satellite TV receiver system, the microwave components are located directly at the antenna feed to avoid losses. A downconverter translates the desired channel information from the 12 GHz broadcast range down to a TV VHF channel, at which losses in the transmission line are more manageable.

A similar arrangement is used in many military radar systems. The rotating antenna with transmitter power amplifier and receive preamplifier can be located in one unstaffed vehicle while the signal processing equipment and sometimes the display systems are located in another at some distance. In this way, the operators and signal processing equipment are not part of the target for an antiradiation missile, especially popular as a counter to air defense radars.

Chapter 2

What Are the Types of Transmission Lines?

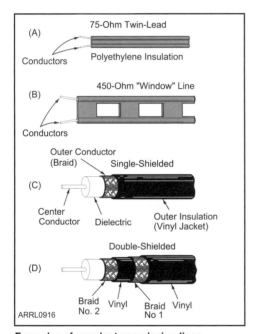

Examples of popular transmission lines.

Transmission lines come in multiple flavors, depending on the application. Here we will broadly categorize lines into their major types. We will also introduce some of the varieties within each category that we will discuss in greater detail in later chapters.

UNBALANCED TRANSMISSION LINE

Transmission lines in which the conductors are at different potentials with respect to ground are referred to as unbalanced transmission lines.

Coaxial Cable

Probably the most frequently encountered type of RF transmission line is coaxial cable. It is found connecting almost every television set to an antenna or to a cable service provider. The cable that cable service providers use is also coaxial cable, at least in the last part of the system, near the subscriber. Coaxial cable is one member of the class of unbalanced transmission lines.

Coaxial cable consists of two conductors, one surrounding the other in a concentric configuration as shown in **Figure 2.1**. In operation, if properly terminated and used, the RF field is contained within the space marked "dielectric" (an insulating material) between the two conductors. At the ends, the outer conductor, also called the *shield*, can be at ground potential. The inner conductor is at a potential above ground, thus the name "unbalanced."

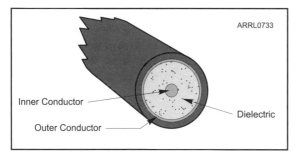

Figure 2.1 – The configuration of coaxial cable. The inner conductor is often wire, sometimes stranded for additional flexibility. Rigid cable may have a center conductor of solid wire or sometimes copper pipe. The outer conductor can be braided copper, wrapped tinfoil, or for rigid cable, copper or aluminum tubing. The lowest-loss rigid coax is made of copper tubing for both conductors with a dielectric mostly of air. Occasional ceramic spacers are employed to maintain the spacing. Other cables have a polyethylene dielectric, or sometimes a foamed poly that acts like something in between.

At RF frequencies, the phenomenon known as *skin effect* keeps the RF shield currents on the inside of the outer conductor. The outside of the outer conductor then acts almost as if it were a separate third conductor at ground potential along its length. This, along with the fact that the net field is within the cable, provides the effect of shielding the signals within the cable from the effect of any signals or other effects outside the cable. This is a major advantage of coaxial cable.

Within the class of coaxial cable ("coax" to its friends), there are a variety of subdistinctions:

♦ Characteristic impedance — The ratio of the conductor diameters and the properties of the dielectric determine the characteristic impedance, a topic that will get a chapter of its own.

♦ Cable diameter — The diameters together may scale up and down from less than ⅛ inch to many inches. The larger sizes are needed for high-power applications, and also have less loss than smaller-sized cable.

♦ Cable flexibility — There are both rigid and flexible coaxial cables, and some semi-rigid, or perhaps semi-flexible, a major factor in selection based on application. The rigid, and some semi-rigid, cables are usually made with a solid tube, rather than a woven outer conductor structure. Solid outer conductors result in coax with the potential for shielding properties closer to ideal than those with woven shields.

♦ Cable dielectric material — The properties of the material between the conductors, the dielectric, is the major factor in the cable loss at higher frequencies and also a factor in determining the cable characteristic impedance.

Unbalanced Microstrip Transmission Line

Imagine slitting the outer conductor along a length of coaxial transmission line and unfolding it so it lies flat. This is the essential configuration of microstrip transmission line. Its major application is found in intraequip-

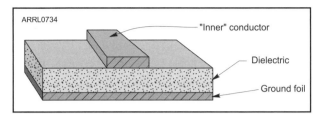

Figure 2.2 – Unbalanced microstrip line. A conductive microstrip left on one side of a piece of two-sided printed circuit board can be operated as an unbalanced transmission line with reference to the foil on the other side. While it acts a lot like coax in some ways, it does not share its shielding properties. In some implementations, a second board or layer of a multilayer board makes a sandwich with additional shielding. In all cases, the connections must be made close to the beginning and end of the conductors. With the sandwich configuration, the ground connection must be made to both ground planes near the location of the start and end of the conductive path.

ment wiring of assemblies made with double-sided printed circuit board. The line can be formed by etching away some of the conductor on one side of the board (see **Figure 2.2**) as part of the process of generating the rest of the wiring on the board.

While microstrip line is efficient from a manufacturing standpoint, it does not offer the shielding qualities of coaxial cable. The fields between the conductors will extend outward from the space between the conductors, requiring care in locating wiring for other parts of the circuit to avoid coupling between circuits. Of course, if coupling is a problem, small coaxial cable is well suited to PC board use, if not quite as easy to implement.

Single Wire Line

Before the days of radio, a single wire was used as a transmission medium for wire-line telegraph systems. The wire was fed against an earth/ground connection and was useful but very lossy. It is not particularly suited for RF systems since it will act more like an antenna, unless it is very close to a highly conductive ground. Think virtual microstrip line.

BALANCED LINE

A transmission line with two parallel conductors can be used as a balanced transmission system — meaning that each conductor is at the same magnitude of potential compared to ground but of opposite sign. This means that at a particular instant, if one wire is at +5 V, for example, the other will be at –5 V, with respect to some common level or ground reference.

Open-Wire Line

The earliest balanced transmission line was a pair of wires at a constant spacing insulated from ground (see chapter photo and **Figure 2.3**). Before the days of radio, and for some time thereafter, such a pair was used for each telephone circuit, making for very imposing utility poles with many cross Ts, almost blocking out the sun in large cities. Current telephone

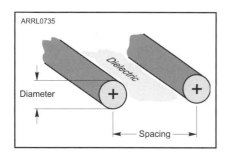

Figure 2.3 – Balanced transmission line. Just two wires, spaced a fixed distance and separated by a dielectric — air, or some other insulating material.

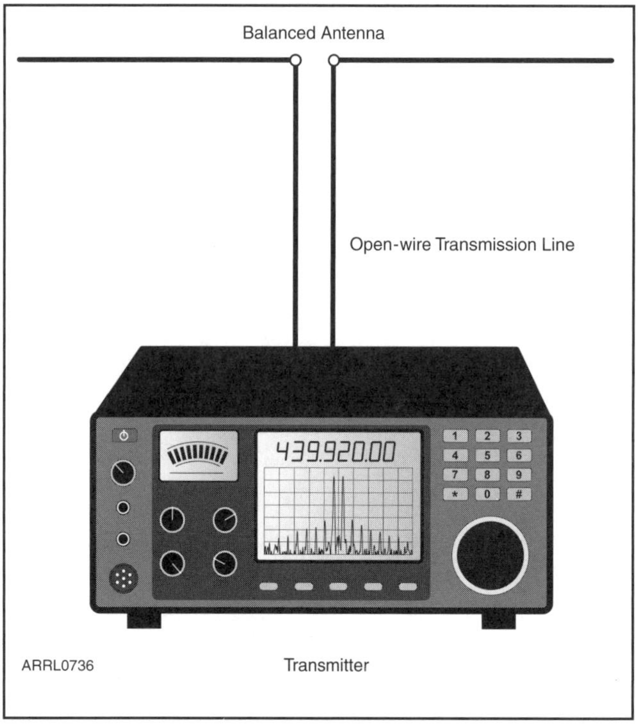

wiring uses insulated twisted pairs, bundled with others in large cables supporting multiple end-user customers. Once individual circuits reach a switching center, they are concentrated with switches and routed between centers on high-density fiber optic links.

Very early radio was conducted on low frequencies typically using large top-loaded monopole antennas fed directly from the transmitter.[1] As operating frequencies became higher, horizontal antennas could be high enough to be useful, and balanced antenna structures, such as a center-fed dipole, became popular. A natural way to interconnect the typical balanced output of the transmitter's tuned transformer to the center of a balanced antenna was to use balanced open-wire line (see **Figure 2.4**).

The wires of open-wire line require some method to hold them at a relatively constant distance. In early telephone systems, this was accomplished by tying each wire to a glass insulator held to a cross T by a peg. In recent rural travel, I saw quite a few of these still in use on poles along railroad right-of-way, likely for signaling. Early radio transmission lines used paraffin-soaked wood dowels to position the wires. These gave way to glazed molded ceramic insulators made for the purpose. Current amateur practice is to use PVC plumbing pipe or acrylic plastic with slits to hold the wires (see **Figure 2.5**).[2]

Figure 2.4 – An open-wire transmission line is a natural for interconnecting a balanced output transmitter and a balanced antenna.

Twinlead

After World War II, television started to become popular in the US. The signals were transmitted over the air using a portion of the VHF range. All antenna designs of the time were of balanced feed and used easily installed transmission line developed for TV applications.

The answer was called twinlead. It was a balanced transmission line made from a pair of stranded wires held about half an inch apart and enclosed in a polyethylene web that insulated each wire and filled the space in between with a flexible insulating material (as shown at the top of the chapter photo). This was the line of choice for this application, as well as some amateur antenna systems until it was displaced in TV service by coaxial cable.

Window Line

This derivative of twinlead is made for higher power and is often used in Amateur Radio transmitting applications. Window line is constructed similarly to twinlead, but has windows cut in the webbing along its length. The windows make the line act more like open-wire line in terms of attenuation, without having the complication of discrete insulators.

The wider spacing typically results in a higher characteristic impedance than regular twinlead.

Figure 2.5 – The production line is in full swing as Barry Shackleford, W6YE, fabricates his own open-wire line. [BARRY SHACKLEFORD, W6YE]

Twisted Pair

A pair of wires twisted together can be used as a makeshift balanced transmission line. The characteristic impedance will be close to 100 Ω. This type of line is not often used for RF applications, although it can serve in a pinch. It is frequently encountered in telephone and local area network (LAN) wiring.

Shielded Twisted Pair

By enclosing a twisted pair in a shield, we can get some of the benefits of balanced line, along with the shielding effect of coax. Of course, we get the disadvantages of each as well, primarily the high attenuation of coax as compared to most balanced lines. The trade name for this cable is Twinax, and it has most recently been used in computer networking, but was once a choice for RF applications as well.

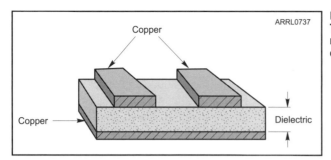

Figure 2.6 – Balanced microstrip line. This works the same way as unbalanced microstrip line, but the two conductors on one side act as a balanced pair.

Balanced Microstrip Line

A pair of etched conductors on PC board material can serve as a balanced line within equipment in the same way that microstrip can provide unbalanced connectivity (see **Figure 2.6**).

Parallel Coax Lines

Two similar coaxial cables of the same length can be operated as a balanced line if the balanced signals are connected to the inner conductors and the shields are tied together at both ends (see **Figure 2.7**). This is not a good choice for many applications of balanced line, since the attenuation is the same as for coax and the cost is twice as high as a single coax. It does work well for balanced line installations in which a small section of the line must be on the ground or in some environment that is unsuitable for open-wire line. Because the fields are entirely within the coax, the two lengths can be run separately as long as they are the same length and the shields at each end are connected together with short connections.

GUIDED WAVE STRUCTURES

UHF and especially microwave (UHF above 1000 MHz) signals are sometimes transported using a kind of mechanism different from a transmission line. The signal loss associated with coaxial cable at UHF and higher frequencies makes them unattractive for all but very short connection paths. A better alternative is to transmit the signals as electromagnetic waves propagating in space. An antenna could make such a transition, but suffers because the signal will spread spatially as it propagates. Instead, the waves are guided by propagating them within a different kind of medium.

The structures described below are a substitute for transmission lines. They are not frequently encountered in Amateur Radio, except when used by some microwave specialists. Beyond the short description below, they will not be covered in this book.

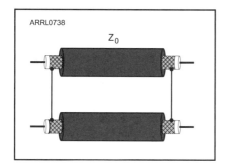

Figure 2.7 – Pair of coaxial cables used as a balanced transmission line. This is particularly useful as a short section of a balanced run to transit an area that is not appropriate for the fields surrounding open-wire line. The attenuation is as high as coax, so short sections are key. It is important that the shields are interconnected at both ends as shown. Since the shields are at reference potential in a balanced system, either may also be connected to structural ground, if appropriate.

Waveguide

The most common guided wave structure is a hollow metal tube called a waveguide. The waves are launched into the tube by an antenna-like probe, propagate within the tube, and are extracted by another probe at the far end. In order to carry the waves without significant attenuation, the guide needs to have a width of at least ½ wavelength and is often of rectangular cross-section. The size restriction limits the usefulness of the technology to the higher UHF frequencies, although I have seen waveguide in use in military radar installations at frequencies as low as 450 MHz — the waveguide tubes looked like very large heating ducts.

Single Wire Transmission Line

It is also possible to guide microwave energy along a wire. This method was originally called Goubau line (after its inventor), or G-line for short. The wire was surrounded by a special dielectric with the appropriate properties for slowing the wave in order to keep it traveling along the wire, rather than radiating as if it were an antenna. Special conical launchers are used at each end of the line, to and from coaxial transmission line. This type of line is easier to deploy in the field than waveguide since it can be rolled up, but it must be used in straight runs. G-line was part of a military microwave communications system that I encountered in the 1960s.

It was later found that waves of a different orientation could be launched onto an uninsulated wire and transmitted with low attenuation using a different configuration of launcher. This method is called E-line.

Notes
[1] J. Hallas, W1ZR, *Basic Antennas*. Available from your ARRL dealer or the ARRL Bookstore, ARRL order no. 9994. Telephone 860-594-0355, or toll-free in the US 888-277-5289; **www.arrl.org/shop**; **pubsales@arrl.org**.
[2] B. Shackleford, W6YE, "Custom Open-wire Line — It's a Snap," *QST*, Jul 2011, pp 33-36.

Chapter 3

Let's Examine Coaxial Transmission Line

Coaxial cables going up one of the towers at W1AW, the ARRL Headquarters station. An advantage of coax is that the runs can be in close proximity to each other without interference.

Coax cable is the most commonly encountered type of transmission line. Coax is composed of two concentric conductors separated by a dielectric material (see **Figure 3.1**). While some types have an exposed outer conductor, the commonly encountered flexible cables and some rigid types are covered by a protective insulating jacket. Coax is found in many sizes for different applications, with the most common ranging from about 0.075 inches to a bit less than half an inch in diameter.

While individual conductors are occasionally connected directly to circuitry, most often within equipment, in most applications the cables are fitted with one of a number of special connectors on their ends. There are different series of connectors, some appropriate for higher frequencies, some waterproof, and some selected just because of the equipment to which they must connect.

EQUIVALENT CIRCUIT OF A COAXIAL TRANSMISSION LINE

Any two conductors separated by a dielectric will have electrical capacitance between the conductors. The capacitance is a function of the surface area of the conductors, the separation between them, and the dielectric constant of the insulation material between them. Coax cable has these properties, with capacitance created by the insulation between the inner surface of the outside conductor and the outer surface of the inside conductor. Any dielectric between the conductors will increase the capacitance compared

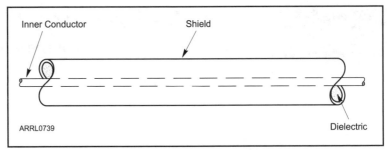

Figure 3.1 – Sketch of a typical coaxial transmission line.

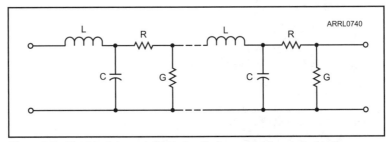

Figure 3.2 – Electrical dc equivalent circuit of a coaxial transmission line. At higher frequencies the series resistance increases because of skin effect. The length of each section can be changed by scaling the values. For many analysis applications, having 10 sections per wavelength will provide reasonable accuracy, thus the one-foot sections might be usable up to around 100 MHz, if the resistance were corrected for skin effect.

to air-insulated cable. The total area of each conductor is a function of the cable length, so the capacitance is usually specified in terms of capacitance per unit length, notated in this country in pF/ft. For example, the usual 50 Ω coax cable (which will be discussed shortly) might have a capacitance of 29 pF/ft.

Similarly, any piece of wire will have a certain inductance, and that includes the conductors making up a coaxial cable. Again, since the inductance is also a function of the cable length, it too is usually expressed in terms of inductance per unit length. For typical lines as above, it will be about 72.5 nH/ft.

The conductors will also have electrical resistance. Because of their relative sizes, this will predominantly be the resistance of the inner conductor. At very low frequencies, the resistance will be around the dc resistance of the inner conductor, perhaps 1.8 mΩ/ft depending on conductor size. There will also be some loss in the dielectric, increasing with frequency. This is represented as the conductance G. We thus could model a transmission line as an electrical circuit such as that shown in **Figure 3.2**.

CHARACTERISTIC IMPEDANCE

The characteristic impedance (Z_0) of a transmission line is one of its most significant parameters. If we consider the line model shown in Figure 3.2 and imagine a dc voltage connected to one end, a current will start to flow, regardless of what is happening at the other end. If the line were just capacitive, the initial current would be very high, as it tried to charge the capacitance. If it were entirely inductive, the initial current would be zero because of the back EMF. Because it is a combination, the initial current will be in between the two values, also limited by the resistance, and not usually a significant portion of the effect. That current will continue until the leading edge of the current pulse gets to the far end and returns with an indication of the effect of the load. The current pulse will travel at somewhat less than the speed of light in free space.

The amount of the current will be the same as if the source were connected to an impedance — generally close to being a resistor — during the initial period. The value of that resistor is called the *characteristic impedance* of the line, usually referred to as its Z_0. If the far end of the line were connected to an actual resistor equal to the Z_0, the same current would continue to flow until the source was disconnected, just as though it were an infinite length of line.

If we use the transmission line for the carrying of an RF signal rather than dc, the same kind of thing happens. For example, using Ohm's law, if we put a 1 V_{RMS} RF signal on one end of a line with a Z_0 of 50 Ω, the initial current would be 1/50 = 0.02 A or 20 mA. If the line were lossless (the R in the model = 0) and terminated at the far end with a 50 Ω resistor, the same 20 mA would flow into the resistor. This is known as a *matched* condition. If the termination were different than a 50 Ω resistor, we could no longer still have 1 V_{RMS} and 20 mA into the load, because it would violate Ohm's law for a different value resistor. Some fraction of the signal would be reflected from the end so that the combination of the forward and reflected signals would result in a voltage and current at the load that satisfies Ohm's law.

The sum of the forward and reflected wave is referred to as a *standing wave*, because the amplitude at any point along the line stays constant, just as it has to stay constant at the termination. This means that the voltage and current along the line will vary, repeating every half wavelength until it is back to the source (as shown in **Figure 3.3**). The impedance at the source will not be the value of the termination nor the line Z_0, but a function of the termination impedance and the distance from it (in wavelengths).

The ratio of maximum voltage (or current) to the minimum voltage (or current) on the line is called the standing wave ratio (SWR), a major consideration in most systems. By having a termination equal to the line

I Know What's Happening at the Shack — What's Happening at the Other End of my Feed Line?

If you want to find out — here's the easy way, using TLW.

Joel R. Hallas, W1ZR
Technical Editor, QST

I'm told that one of the more frequent questions received by QST's "Doctor" has to do with folks wanting to determine the impact of transmission line losses on the effectiveness of their antenna system. These questions are often along the lines of, "I measure an SWR of 2.5:1 at the transmitter end of 135 feet of RG-8X coaxial cable. My transceiver's auto-tuner can tune it to 1.1, but how can I tell what my losses are?" or "How much difference will I have if I have a tuner at the antenna instead of using the built-in tuner?"

These are important questions that almost every amateur operator is faced with from time to time. An approximate answer can be obtained by using the graphs found in any recent edition of *The ARRL Antenna Book* showing the loss characteristics of many transmission line types, plus adding in the effect of an SWR greater than 1:1. The SWR at the antenna end can be determined from the bottom end SWR and the cable loss. Using these graphs requires a bit of interpolation or Kentucky windage, but can result in useful data.

But There's an Even Better Way!

Packaged with each of the last few editions of *The ARRL Antenna Book* is a CD containing the pages of the whole *Antenna Book* as well as some very useful software. The program that I use almost daily is one written by *Antenna Book* Editor R. Dean Straw, N6BV, called *TLW*, for *Transmission Line for Windows*.

TLW provides a very easy-to-operate mechanism to determine everything I usually need to know about what's happening on a transmission line. When you open the program, you are presented with a screen as shown in **Figure A**. This has the values plugged in from the last time you used it, often saving a step. Let's take a quick tour of the inputs:

Cable Type — This allows you to select the cable you would like to analyze. A drop-down

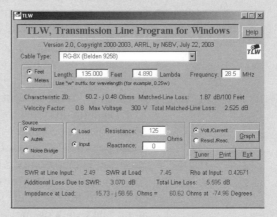

Figure A — The opening screen of *TLW*, illustrating the process described in the article.

box provides for the selection of one of 32 of the most common types of coax and balanced lines. An additional entry is provided for User Defined Transmission Lines that can be specified by propagation velocity and attenuation.

Length — In feet or meters, your choice.

Frequency — This is an important parameter when dealing with transmission line effects.

Source — This defines the form of the input impedance data. Generally, you can use NORMAL.

Impedance — The impedance can be specified as what you measure, resistive (real) and reactive (imaginary, minus means capacitive). This could come from your antenna analyzer at either end of the transmission line. Note, if you only know the SWR, not the actual impedance, all is not lost — see below.

Now for the Outputs

SWR — The SWR is provided at each end of the cable. This is an important difference that many people miss, important even with a moderate SWR at the transmitter end, as we'll see — the SWR at the antenna will be much higher due to the cable loss. With *TLW*, you instantly know the SWR at both ends, and the loss in the cable itself.

Rho at Load — This is the reflection coefficient, the fraction of the power reflected back from the load.

Additional Loss Due to SWR — This is one of the answers we were after.

Total Loss — And this is the other, the total loss in the line, including that caused by the mismatch.

But Doctor, What if I can Only Measure the SWR — Not the Actual Impedance?

Often the only measurement data available is the SWR at the transmitter end of the cable. Because the losses are a function of the SWR, not the particular impedance, you can just put in an arbitrary impedance with that same SWR and click the INPUT button. An easy arbitrary impedance to use is just the SWR times the Z0 of the cable, usually 50 Ω. For example, you could use a resistance of 125 Ω to represent an SWR of 2.5:1. This is what we've done in Figure A, using 135 feet of popular Belden RG-8X.

The results are interesting. Note that the 2.5:1 SWR as seen at the radio on 28.5 MHz results from a 7.45:1 SWR at the antenna — perhaps this is an eye-opener! Note that of the 5.6 dB loss, more than half, or 3.1 dB, is due to the mismatch. Note that if we used something other the actual measured impedance, we can't make use of the impedance data that *TLW* provides. We can use the SWR and loss data, however, but that's probably what we wanted to find out.

We can now do some "what ifs." We can see how much loss we have on other bands by just changing the frequency. For example, on 80 meters, with the same 2.5:1 at the transmitter end, the SWR at the antenna is about 3:1 and the loss is slightly more than 1 dB. We could also plug in an impedance calculated at the antenna end and see what difference other cable types would make. For example, with the same 28.5 MHz SWR of 7.45 at the antenna and 135 feet of ½ inch Andrew Heliax, we will have a total loss of 1.5 dB at 28.5 MHz. Note that the SWR seen at the bottom will now be 5.5:1 and our radio's auto-tuner might not be able to match the new load.

But Wait, There's More!

You can also click the GRAPH button and get a plot of either voltage and current or resistance and reactance along the cable. Note that these will only be useful if we have started with actual impedance, rather than SWR.

Pushing the TUNER button results in a page asking you to select some specifications for your tuner parts. *TLW* effectively designs a tuner of the type you asked for at the shack end of the cable. It also calculates the power lost in the tuner and gives a summary of the transmitted and lost power in watts, so you don't need to calculate it!

When you've finished, be sure to hit the EXIT button; don't just close the window. Otherwise *TLW* may not start properly the next time you want to use it.

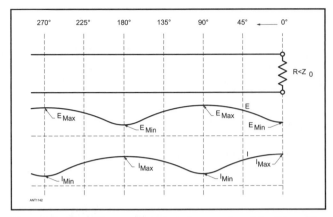

Figure 3.3 – The voltage and current along a mismatched line will vary, repeating every half wavelength.

Z_0, the standing wave doesn't exist and the SWR is 1:1. For a number of reasons, this is a desirable condition and cable with different values of Z_0 are available, including most commonly 35 Ω, 50 Ω, 75 Ω, and 93 Ω.

FIELDS WITHIN COAX

We need to briefly consider both the electric and magnetic fields in coax to understand how and why it works as it does. The electric field, just as with the plates of any capacitor with an applied voltage, exists between the outside of the center conductor and the inside of the outer conductor (see **Figure 3.4**). Because the outer conductor surrounds the inner conductor, all the electric field terminates on the inside of the shield and does not couple to, nor is it affected by, nearby conductors. Thus, the outside of the cable is effectively shielded from the effects of the electric field within the cable — one of the major benefits of coax.

As in any conductor, a magnetic field surrounds the two current-carrying conductors. For properly operating coaxial cable, the two currents will be equal and opposite; thus, the fields will cancel in the region outside the cable (see **Figure 3.5**), so the net magnetic field surrounding the cable will be zero. This, combined with the shielding of the electric field, results in a transmission line that is independent of the electrical conditions of its surroundings. This means it can be coiled, bundled with other cables, or run within metal conduit without any problems — as long as the currents are balanced and the shield is solid.

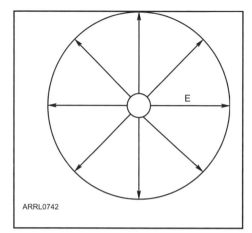

Figure 3.4 – The electric field within a coaxial transmission line. Note that if the shield is solid, it exists only within the cable, resulting in a part of the shielding effect.

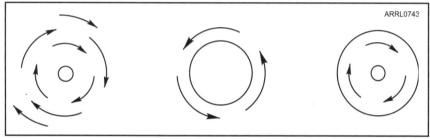

Figure 3.5 – As in any conductor, the magnetic field surrounds the two current-carrying conductors. On the left, the magnetic field surrounding a single isolated conductor is shown. The center image shows the magnetic field surrounding the outer conductor, and at right is shown the magnetic field surrounding the inner conductor. For properly operating coaxial cable, the two currents will be equal and opposite, as shown, and thus the fields will cancel outside the cable.

SHIELDING LIMITATIONS OF REAL COAX

Coax in the real world can be quite close to the ideal in most respects, but there are some limitations. The electric field shielding effectiveness is a function of the shield coverage. The best shielded cables are those with a solid tubular outer conductor. These can be considered a 100% shield, although loose or improperly installed connectors can be leakage points.

Unfortunately, solid tubular coax is not applicable to the many applications that require flexible cable. Probably the next best type, in terms of shielding effectiveness, is the type used by the cable television industry. It usually has multiple shields, generally one of wrapped aluminum foil and another of braided tinned copper. This is also quite effective — and has to be, otherwise cable signals, which occupy most of the radio spectrum, would interfere with other users, including such critical systems as aircraft communications and navigation.

The rest of the coax world uses braided copper shielding. This can be adequate for most applications, although the quality varies considerably among brands and types. In many cases the manufacturer will specify the shielding coverage, usually in terms of a percentage, with 100% meaning fully covered. Some cable types have double shields — two independent braided shields, one around the other, in electrical contact over their length. These can be very close to 100% effective. Less expensive cables can have effectiveness as low as in the 80% range. The less completely shielded cables allow a partial leakage of the electric field, allowing some coupling to and from nearby circuits.

Another coupling mechanism occurs if the shield current is different from the current on the inner conductor. This can occur in a number of ways, typically due to a termination in which the outside of the shield acts like a separate conductor from the inside of the shield. Currents on the outside can also be induced by proximity to an antenna system.

TYPES OF COAXIAL CABLE

Coaxial cable came into common use during World War II, mainly in military radar and radio applications. A set of standards was developed during the period for radio guide (RG) cable, with that designation followed by a series number. The most common in radio use over the years have been 50 Ω characteristic Z_0 RG-8 and RG-58. Each has a copper center conductor, a polyethylene dielectric, a braided copper shield, and an outer jacket, originally made from black vinyl but more often of PVC or polyethylene today. The primary difference between the two is size, with the RG-8 having an outside diameter of 0.405 and RG-58 having one of 0.195 inches. The smaller cable has higher attenuation and lower voltage and power ratings, but offers lighter weight, additional flexibility, and handling convenience.

The next most frequently encountered early cables were those with 75 Ω characteristic impedance, most commonly RG-11 at 0.405 inches and RG-59 at 0.242 inches. A suffix "U" indicates cable that is universal (usable for multiple functions), while other suffixes indicate special characteristics or construction differences. Probably the most common coax in use today is RG-6 type (see below), an RG-59-size 75 Ω cable with a more effective foil shield, used for in-house connection of cable television equipment.

These characteristics were documented in a US military procurement manual that was cancelled in 2001, so the early numbers are no longer subject to official standardization. Thus cable that is called "RG-8U type"

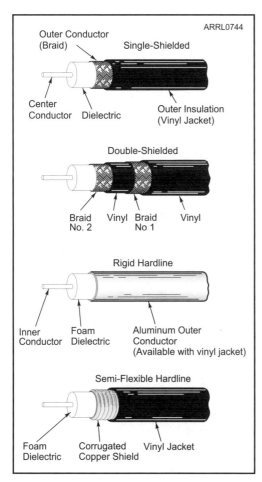

Figure 3.6 – Construction of a number of common types of coaxial cable.

is a manufacturer's designation that may mean the cable is somewhat like RG-8, but does not guarantee that it meets either early or current military standards. It may have a less dense shield or lower quality jacket, for example. Later standards include MIL-DTL-17H from August 19, 2005, which superseded MIL-C-17G of March 9, 1990. These standards refer to newer standard cable series that have replaced the earlier types. For example, today one would obtain RG-223 instead of RG-58, and RG-213 instead of RG-8. As in the earlier series, the word "type" is a manufacturer's escape clause that allows them to skirt military specifications.

Table 3.1 is a listing of many of the types of coaxial transmission line, as well as other types that we will discuss later. Note that within a type of cable, there are multiple cables from different manufacturers, with quite different characteristics. Some have a foamed polyethylene dielectric (FPE) instead of solid polyethylene. FPE is partly air and thus has a lower relative dielectric constant and less loss than standard cable. The foam also makes for a less sturdy cable that is more susceptible to water migration and cannot be flexed as tightly as the solid dielectric coax. The construction of a number of coaxial cable types is shown in **Figure 3.6**.

KEY CABLE PARAMETERS

The headings in Table 3.1 give insight into the differences between cable types. Each is described briefly below.

Nominal Z_0 (Ω) — As was discussed earlier, Z_0 is a primary transmission line parameter. Most systems specify the design impedance for various connections, and typically that will align with a standard cable Z_0. The usual practice is to use a cable with that Z_0 to connect to equipment to a load with the same impedance, although we will discuss exceptions. In most cases, particularly at lower frequencies, the difference between a Z_0 of 50, 51, 52, or 53 Ω, as shown in the RG-58 types, will not be significant. For coaxial cable, the Z_0 can be determined by the dimensions as follows:

Table 3.1
Nominal Characteristics of Commonly Used Transmission Lines

G or Type	Part Number	Nom. Z_0 Ω	VF %	Cap. pF/ft	Cent. Cond. AWG	Diel. Type	Shield Type	Jacket Material	OD inches	Max V (RMS)	Matched Loss (dB/100') 1 MHz	10	100	1000
RG-6	Belden 1694A	75	82	16.2	#18 Solid BC	FPE	FC	P1	0.275	600	0.2	0.7	1.8	5.9
RG-6	Belden 8215	75	66	20.5	#21 Solid CCS	PE	D	PE	0.332	2700	0.4	0.8	2.7	9.8
RG-8	Belden 7810A	50	86	23.0	#10 Solid BC	FPE	FC	PE	0.405	600	0.1	0.4	1.2	4.0
RG-8	TMS LMR400	50	85	23.9	#10 Solid CCA	FPE	FC	PE	0.405	600	0.1	0.4	1.3	4.1
RG-8	Belden 9913	50	84	24.6	#10 Solid BC	ASPE	FC	P1	0.405	600	0.1	0.4	1.3	4.5
RG-8	CXP1318FX	50	84	24.0	#10 Flex BC	FPE	FC	P2N	0.405	600	0.2	0.6	1.5	4.5
RG-8	Belden 9913F7	50	83	24.6	#11 Flex BC	FPE	FC	P1	0.405	600	0.2	0.5	1.5	4.8
RG-8	Belden 9914	50	82	24.8	#10 Solid BC	FPE	FC	P1	0.405	600	0.1	0.4	1.4	4.8
RG-8	TMS LMR400UF	50	85	23.9	#10 Flex BC	FPE	FC	PE	0.405	600	0.1	0.5	1.6	4.9
RG-8	DRF-BF	50	84	24.5	#9.5 Flex BC	FPE	FC	P2N	0.405	600	0.1	0.5	1.6	5.2
RG-8	WM CQ106	50	84	24.5	#9.5 Flex BC	FPE	FC	P1	0.405	600	0.2	0.6	1.8	5.3
RG-8	CXP008	50	78	26.0	#13 Flex BC	FPE	S	P1	0.405	600	0.1	0.5	1.8	7.1
RG-8	Belden 8237	52	66	29.5	#13 Flex BC	PE	S	P1	0.405	3700	0.2	0.6	1.9	7.4
RG-8X	Belden 7808A	50	86	23.5	#15 Solid BC	FPE	FC	PE	0.240	600	0.2	0.7	2.3	7.4
RG-8X	TMS LMR240	50	84	24.2	#15 Solid BC	FPE	FC	P1	0.242	300	0.2	0.8	2.5	8.0
RG-8X	WM CQ118	50	82	25.0	#16 Flex BC	FPE	FC	P2N	0.242	300	0.3	0.9	2.8	8.4
RG-8X	TMS LMR240UF	50	84	24.2	#15 Flex BC	FPE	FC	P1	0.242	300	0.2	0.8	2.8	9.6
RG-8X	Belden 9258	50	82	24.8	#16 Flex BC	FPE	S	P1	0.242	600	0.3	0.9	3.1	11.2
RG-8X	CXP08XB	50	80	25.3	#16 Flex BC	FPE	S	P1	0.242	300	0.3	0.9	3.1	14.0
RG-9	Belden 8242	51	66	30.0	#13 Flex SPC	PE	SCBC	P2N	0.420	5000	0.2	0.6	2.1	8.2
RG-11	Belden 8213	75	84	16.1	#14 Solid BC	FPE	S	PE	0.405	600	0.2	0.4	1.3	5.2
RG-11	Belden 8238	75	66	20.5	#18 Flex TC	FPE	S	P1	0.405	600	0.2	0.7	2.0	7.1
RG-58	Belden 7807A	50	85	23.7	#18 Solid BC	FPE	FC	PE	0.195	300	0.3	1.0	3.0	9.7
RG-58	TMS LMR200	50	83	24.5	#17 Solid BC	FPE	FC	PE	0.195	300	0.3	1.0	3.2	10.5
RG-58	WM CQ124	52	66	28.5	#20 Solid BC	PE	S	PE	0.195	1400	0.4	1.3	4.3	14.3
RG-58	Belden 8240	52	66	28.5	#20 Solid BC	PE	S	P1	0.193	1900	0.3	1.1	3.8	14.5
RG-58A	Belden 8219	53	73	26.5	#20 Flex TC	FPE	S	P1	0.195	300	0.4	1.3	4.5	18.1
RG-58C	Belden 8262	50	66	30.8	#20 Flex TC	PE	S	P2N	0.195	1400	0.4	1.4	4.9	21.5
RG-58A	Belden 8259	50	66	30.8	#20 Flex TC	PE	S	P1	0.192	1900	0.4	1.5	5.4	22.8

G or Type	Part Number	Nom. Z_0 Ω	VF %	Cap. pF/ft	Cent. Cond. AWG	Diel. Type	Shield Type	Jacket Material	OD inches	Max V (RMS)	Matched Loss (dB/100') 1 MHz	10	100	1000
RG-59	Belden 1426A	75	83	16.3	#20 Solid BC	FPE	S	P1	0.242	300	0.3	0.9	2.6	8.5
RG-59	CXP 0815	75	82	16.2	#20 Solid BC	FPE	S	P1	0.232	300	0.5	0.9	2.2	9.1
RG-59	Belden 8212	75	78	17.3	#20 Solid CCS	FPE	S	P1	0.242	300	0.6	1.0	3.0	10.9
RG-59	Belden 8241	75	66	20.4	#23 Solid CCS	PE	S	P1	0.242	1700	0.6	1.1	3.4	12.0
RG-62A	Belden 9269	93	84	13.5	#22 Solid CCS	ASPE	S	P1	0.240	750	0.3	0.9	2.7	8.7
RG-62B	Belden 8255	93	84	13.5	#24 Flex CCS	ASPE	S	P2N	0.242	750	0.3	0.9	2.9	11.0
RG-63B	Belden 9857	125	84	9.7	#22 Solid CCS	ASPE	S	P2N	0.405	750	0.2	0.5	1.5	5.8
RG-142	CXP 183242	50	69.5	29.4	#19 Solid SCCS	TFE	D	FEP	0.195	1900	0.3	1.1	3.8	12.8
RG-142B	Belden 83242	50	69.5	29.0	#19 Solid SCCS	TFE	D	TFE	0.195	1400	0.3	1.1	3.9	13.5
RG-174	Belden 7805R	50	73.5	26.2	#25 Solid BC	FPE	FC	P1	0.110	300	0.6	2.0	6.5	21.3
RG-174	Belden 8216	50	66	30.8	#26 Flex CCS	PE	S	P1	0.110	1100	1.9	3.3	8.4	34.0
RG-213	Belden 8267	50	66	30.8	#13 Flex BC	PE	S	P2N	0.405	3700	0.2	0.6	1.9	8.0
RG-213	CXP213	50	66	30.8	#13 Flex BC	PE	S	P2N	0.405	600	0.2	0.6	2.0	8.2
RG-214	Belden 8268	50	66	30.8	#13 Flex SPC	PE	D	P2N	0.425	3700	0.2	0.6	1.9	8.0
RG-216	Belden 9850	75	66	20.5	#18 Flex TC	PE	D	P2N	0.425	3700	0.2	0.7	2.0	7.1
RG-217	WM CQ217F	50	66	30.8	#10 Flex BC	PE	D	PE	0.545	7000	0.1	0.4	1.4	5.2
RG-217	M17/78-RG217	50	66	30.8	#10 Solid BC	PE	D	P2N	0.545	7000	0.1	0.4	1.4	5.2
RG-218	M17/79-RG218	50	66	29.5	#4.5 Solid BC	PE	S	P2N	0.870	11000	0.1	0.2	0.8	3.4
RG-223	Belden 9273	50	66	30.8	#19 Solid SPC	PE	D	P2N	0.212	1400	0.4	1.2	4.1	14.5
RG-303	Belden 84303	50	69.5	29.0	#18 Solid SCCS	TFE	S	TFE	0.170	1400	0.3	1.1	3.9	13.5
RG-316	CXP TJ1316	50	69.5	29.4	#26 Flex BC	TFE	S	FEP	0.098	1200	1.2	2.7	8.0	26.1
RG-316	Belden 84316	50	69.5	29.0	#26 Flex SCCS	TFE	S	FEP	0.096	900	1.2	2.7	8.3	29.0
RG-393	M17/127-RG393	50	69.5	29.4	#12 Flex SPC	TFE	D	FEP	0.390	5000	0.2	0.5	1.7	6.1
RG-400	M17/128-RG400	50	69.5	29.4	#20 Flex SPC	TFE	S	FEP	0.195	1400	0.4	1.1	3.9	13.2
LMR500	TMS LMR500UF	50	85	23.9	#7 Flex BC	FPE	FC	PE	0.500	2500	0.1	0.4	1.2	4.0
LMR500	TMS LMR500	50	85	23.9	#7 Solid CCA	FPE	FC	PE	0.500	2500	0.1	0.3	0.9	3.3
LMR600	TMS LMR600	50	86	23.4	#5.5 Solid CCA	FPE	FC	PE	0.590	4000	0.1	0.2	0.8	2.7
LMR600	TMS LMR600UF	50	86	23.4	#5.5 Flex BC	FPE	FC	PE	0.590	4000	0.1	0.2	0.8	2.7
LMR1200	TMS LMR1200	50	88	23.1	#0 Copper Tube	FPE	FC	PE	1.200	4500	0.04	0.1	0.4	1.3

G or Type	Part Number	Nom. Z_0 Ω	VF %	Cap. pF/ft	Cent. Cond. AWG	Diel. Type	Shield Type	Jacket Material	OD inches	Max V (RMS)	Matched Loss (dB/100') 1 MHz	10	100	1000
Hardline														
½"	CATV Hardline	50	81	25.0	#5.5 BC	FPE	SM	none	0.500	2500	0.05	0.2	0.8	3.2
½"	CATV Hardline	75	81	16.7	#11.5 BC	FPE	SM	none	0.500	2500	0.1	0.2	0.8	3.2
⅞"	CATV Hardline	50	81	25.0	#1 BC	FPE	SM	none	0.875	4000	0.03	0.1	0.6	2.9
⅞"	CATV Hardline	75	81	16.7	#5.5 BC	FPE	SM	none	0.875	4000	0.03	0.1	0.6	2.9
	LDF4-50A Heliax – ½"	50	88	25.9	#5 Solid BC	FPE	CC	PE	0.630	1400	0.05	0.2	0.6	2.4
	LDF5-50A Heliax – ⅞"	50	88	25.9	0.355" BC	FPE	CC	PE	1.090	2100	0.03	0.10	0.4	1.3
	LDF6-50A Heliax – 1¼"	50	88	25.9	0.516" BC	FPE	CC	PE	1.550	3200	0.02	0.08	0.3	1.1
Parallel Lines														
	TV Twinlead (Belden 9085)	300	80	4.5	#22 Flex CCS	PE	none	P1	0.400	**	0.1	0.3	1.4	5.9
	Twinlead (Belden 8225)	300	80	4.4	#20 Flex BC	PE	none	P1	0.400	8000	0.1	0.2	1.1	4.8
	Generic Window Line	405	91	2.5	#18 Solid CCS	PE	none	P1	1.000	10000	0.02	0.08	0.3	1.1
	WM CQ 554	420	91	2.7	#14 Flex CCS	PE	none	P1	1.000	10000	0.02	0.08	0.3	1.1
	WM CQ 552	440	91	2.5	#16 Flex CCS	PE	none	P1	1.000	10000	0.02	0.08	0.3	1.1
	WM CQ 553	450	91	2.5	#18 Flex CCS	PE	none	P1	1.000	10000	0.02	0.08	0.3	1.1
	WM CQ 551	450	91	2.5	#18 Solid CCS	PE	none	P1	1.000	12000	0.02	0.08	0.3	1.1
	Open-Wire Line	600	92	1.1	#12 BC	none	none	none	**		0.02	0.06	0.2	0.7

Approximate Power Handling Capability (1:1 SWR, 40°C Ambient):

	1.8 MHz	7	14	30	50	150	220	450	1 GHz
RG-58 Style	1350	700	500	350	250	150	120	100	50
RG-59 Style	2300	1100	800	550	400	250	200	130	90
RG-8X Style	1830	840	560	360	270	145	115	80	50

Legend:

**	Not Available or varies		FC	Foil + Tinned Copper Braid
ASPE	Air Spaced Polyethylene		FEP	Teflon® Type IX
BC	Bare Copper		Flex	Flexible Stranded Wire
CC	Corrugated Copper		FPE	Foamed Polyethylene
CCA	Copper Cover Aluminum		Heliax	Andrew Corp Heliax
CCS	Copper Covered Steel		N	Non-Contaminating
CXP	Cable X-Perts, Inc.		P1	PVC, Class 1
D	Double Copper Braids		P2	PVC, Class 2
DRF	Davis RF		PE	Polyethylene

S	Single Braided Shield
SC	Silver Coated Braid
SCCS	Silver Plated Copper Coated Steel
SM	Smooth Aluminum
SPC	Silver Plated Copper
TC	Tinned Copper
TFE	Teflon® Systems
UF	Ultra Flex
WM	Wireman

$$Z_0 = \left(\frac{138}{\sqrt{\varepsilon_r}}\right) \times \log_{10}\left(\frac{D}{d}\right)$$

Where D and d are the diameters of the shield and center conductor respectively in the same units. ε_r is the relative dielectric constant of the dielectric (for air, $\varepsilon_r = 1$, for polyethylene $\varepsilon_r = 2.26$).

VF (%) — This is the velocity factor as a percentage of the speed of light in free space (300,000,000 m/s). This is largely a function of the relative dielectric constant of the insulation material. For polyethylene dielectric, the VF is usually 66%, in the 80th percentile for foamed poly, depending on density, and close to 100% for air-insulated lines. In many applications, the velocity is not particularly important, with two notable exceptions — it is critical if lines are used for delay or phasing lines, and it also directly affects the required length of lines used as impedance transformers (covered later). The velocity factor can be found as shown below.

$VF = 100/\sqrt{\varepsilon_r}$

Cap (pF/ft) — This is the capacitance between the inner and outer conductors of the cable. It is a function of the ratio of the diameters and the dielectric constant. Note that the ratio of diameters also determines the Z_0, so it's not surprising that cables of the same Z_0 with similar dielectrics have the same capacitance. Sometimes pieces of transmission line are used as capacitors in circuits; otherwise this is not often important.

Dimensions and Materials — These are often major considerations. It is critical to know the dimensions if determining connector or adapter types. The abbreviations are listed at the bottom of the table. Note that some center conductors are indicated as being flexible — a prime attribute in many applications.

Max V (RMS) — This is the maximum Root Mean Square, or effective voltage that can be safely applied to the cable. Note that this doesn't equate to power, which is listed separately at the bottom and applies for a matched (SWR = 1:1) condition. This voltage might occur briefly for any reason (the maximum peak voltage is 1.4 times the listed voltage), and above that, arcing is possible. The power is generally more of a thermal issue. For example, the listed voltage for RG-58, 300 V_{RMS}, would translate to a power of 1800 W in a 50 Ω system, while the maximum power (at 1.8 MHz) is listed as 1350 W.

Matched Loss (dB/100 feet) — This is the attenuation of 100 feet of the cable when matched to its Z_0. Note that the cable loss increases with frequency and is shown for four frequencies across the spectrum. The loss is linearly related to length, so 200 feet has twice the loss, and 50 feet has half the loss. This is a key differentiator between cable types. Keep in mind that a 1 dB loss represents about a 20% reduction in power, a 3 dB loss is a 50% reduction in power, and 10 dB is a 90% reduction. All things being equal, larger cables have less loss (due to less conductor resistance) as do cables with higher velocity (less dielectric loss). The matched loss of major cable types is shown graphically in **Figure 3.7**.

ADDITIONAL LOSS DUE TO MISMATCH

You may have noticed that all the loss figures stated in Table 3.1 and shown in Figure 3.7 were for the case of a transmission line terminated in its Z_0. The additional loss due to mismatch is a function of both amount of the mismatch (SWR) and the loss if matched. **Figure 3.8** shows the additional loss in dB that occurs as a result of a line not being matched.

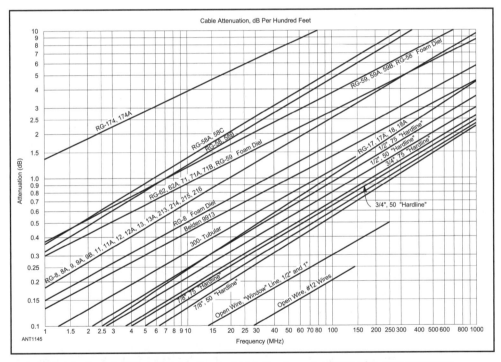

Figure 3.7 – Matched nominal loss of various transmission line types in dB/100 feet of representative transmission line types. As you choose transmission line, it is important to note that this data is representative — the variation between attenuation of different manufacturers and even different part numbers from the same manufacturer can be striking. Check the manufacturer's website for the data sheet of the transmission line you are considering to be sure you know what you are getting. For lengths other than 100 feet, the loss scales linearly. For example, if a line has a matched loss of 2 dB at 100 feet, the same line at the same frequency will have a loss of 1 dB if 50 feet long, and 5 dB if 250 feet long.

Note that the SWR shown is the SWR as measured or calculated at the load, and not the SWR measured at the transmitter end of the cable. This is particularly important in the case of a lossy line, since the loss will reduce both the power reaching the antenna and the power of the reflected wave that is used to determine the SWR. This can give very optimistic and erroneous results.

To give an example of this effect, consider a 100 W transmitter driving 100 feet of coax with a loss of 3 dB (50% loss). The antenna will see 50 W of power. Let's say 20% of the power is reflected due to the antenna mismatch. That will result in 10 W being reflected back toward the source. The 3 dB loss results in 5 W showing up as reflected power at the bottom of the cable. This is quite different than if the 100 W were applied to the antenna on lossless line — in

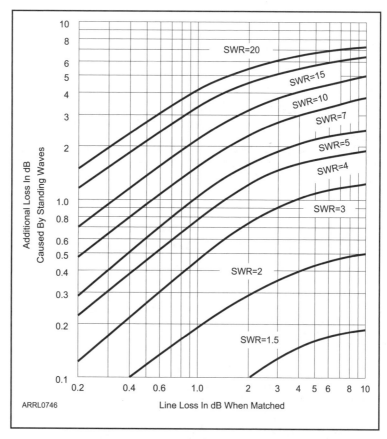

Figure 3.8 – Additional loss of a transmission line when mismatched. This loss needs to be added to the loss in Figure 3.7 for mismatched lines.

Table 3.2
Forward and Reflected Power and SWR as Seen at Each End of a Transmission Line With 3 dB Loss

Measurement	Bottom of Cable	Top of Cable
Forward Power (W)	100	50
Reflected Power (W)	5	10
Indicated Reflection Coefficient	0.224	0.447
Indicated SWR	1.6	4.0

that case, a reflected power of 20 W, not 5 W, would show up at the SWR measurement device.

Table 3.2 summarizes what we have at the two locations. Note the rather distressing result. A very acceptable measurement of an SWR of 1.6:1 at the bottom of the coax is the result of an unpleasant SWR of 4:1 at the antenna. In this example, our 100 W of power results in only 40 W radiated from the antenna — yet all of our measurements make us think we're doing well. Unfortunately, this example is not unusual, especially at the upper end of HF into the VHF range. If it happens at higher frequencies, it is usually more evident since nothing much ends up going in or out of the system! The sidebar discusses ways that this can be calculated, and perhaps avoided, through the use of software.[3]

Notes

[1] J. Hallas, W1ZR, *Basic Antennas*. Available from your ARRL dealer or the ARRL Bookstore, ARRL order no. 9994. Telephone 860-594-0355, or toll-free in the US 888-277-5289; **www.arrl.org/shop**; **pubsales@arrl.org**.

[2] B. Shackleford, W6YE, "Custom Open-wire Line — It's a Snap," *QST*, Jul 2011, pp 33-36.

[3] J. Hallas, W1ZR, "I Know What's Happening at the Shack — What's Happening at the Other End of My Feed Line?" *QST*, Feb 2007, p 63.

Chapter 4

Other Types of Unbalanced Transmission Line

While coaxial cable is the most frequently encountered type of unbalanced transmission line, it is not the only type. This chapter will briefly introduce some other types of line in that category.

MICROSTRIP TRANSMISSION LINE

Microstrip is a form of transmission line that takes advantage of the properties of printed circuit (PC) boards. PC assemblies are a common and efficient method of circuit fabrication that has largely displaced the hand wiring of most electronic systems. The technique employs a thin piece of insulating material, the *substrate*, that serves as the mechanical strength member of the assembly and that supports all the components. Professional and military equipment generally makes use of a fiberglass-based material, while less critical consumer products are often made of phenolic resin.

This microwave circulator is constructed of microstrip transmission line. Microstrip is easily implemented as part of PC board fabrication. The lines shown are single-sided stripline with the foil on the other side of the board serving as the "shield."

A thin layer of copper that will eventually be transformed into component interconnections is bonded to the substrate. More common double-sided PC board has copper on both sides of the substrate. The PC designer (a person or software), working from the schematic of the circuit, lays out the board by specifying where each component will be placed and which pins need to be interconnected. The interconnections are made by removing

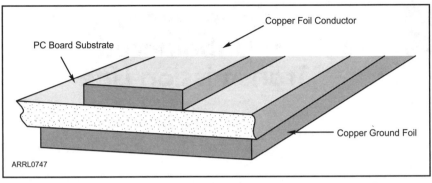

Figure 4.1 – Sketch of a typical microstrip transmission line as etched onto a PC board.

the copper that isn't needed. Often connections are established on both sides of double-sided board, but sometimes one side is used as a ground reference across the board. The unneeded copper is removed by using one of a number of methods of applying "resist" material to the areas that need to be kept. The board is then submerged in an acid bath to remove the excess copper. Following etching, the resist material is removed and the components are mounted to complete the circuit. While this sounds like an arduous process, compared to hand-wiring, it is easily automated and very efficient, especially if multiple boards are required for a production run.

RF circuits within a board, or leaving the board, can be interconnected using miniature coaxial cable with ends connected through holes in the board to the appropriate termination points just as other components with wire leads can be connected. Alternately, a microstrip transmission line can be formed on double-sided board during the printed circuit process. The line must be fabricated above a region that has a ground area on the bottom foil of the board, and merely consists of a conductor etched above the ground foil interconnecting to two end points. This is shown in **Figure 4.1**, along with the electric field between the conductors.

Note that unlike the case of the fields within a coax cable, the fields, while concentrated between the conductors, are not entirely contained there. This is a major limitation of microstrip line — it is not totally shielded. Therefore, care should be taken to make sure that coupling does not occur to sensitive circuits. This can usually be managed at the time of board layout.

In operation it is important that the connections of both input and output ends of the line include direct connections to the end of the "center conductor" and the ground plane directly beneath it.

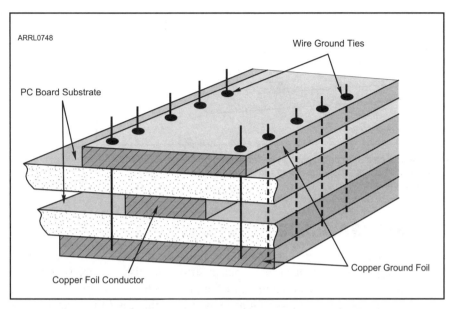

Figure 4.2 – Stripline transmission line, adding another ground layer to microstrip. Note the periodic connections between the two ground strips. These can improve the isolation to close to that of coax.

Characteristic Impedance

The characteristic impedance (Z_0) of microstrip line and the effective dielectric constant of the insulation are determined by the sizes of the conductors and the material used as the dielectric in a manner similar to coaxial cable.

STRIPLINE TRANSMISSION LINE

A variation of microstrip that actually was developed first is called stripline. It is just a microstrip "sandwich," formed by adding another piece of board with a ground foil above the single board in microstrip. Alternately, it can be formed as part of a multilayer board. It can have additional isolation from nearby conductors since the fields are more tightly constrained. It can approach coax in this regard if the two shield foils are tied together along the edge with frequent wires, as shown in **Figure 4.2**. In this case, the ground side of each end termination must connect directly to both ground foils in the area near the inner conductor.

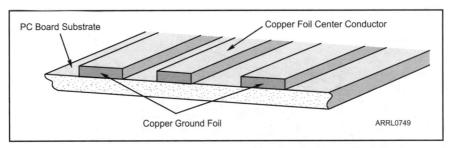

Figure 4.3 – Coplaner strip transmission line. This can be fabricated on single-sided PC board material. The two ground strips must be bonded together at the ends and can be much larger than the center conductor.

COPLANER STRIP TRANSMISSION LINE

Less frequently encountered is a line that can be deployed on a piece of single-sided PC material. Coplaner strip transmission line, as shown in **Figure 4.3**, is not as effective in isolating circuits as the other types, but may be usable in some situations. Note that in this case, the dielectric is a mix of the board substrate below and air above. Again, the ground side of the connection at each end must connect to both ground strips. This can be built into the PC board etching process.

SINGLE WIRE OVER A GROUND PLANE

An insulated wire located close to a metal chassis, or even a piece of PC material with a ground foil, can act like a microstrip line. (See **Figure 4.4**.) This is commonly employed in equipment that is hand-wired in the traditional manner between components on a metal chassis — in many cases without the builder thinking in terms of transmission lines. The effectiveness can be increased by locating the wire in a corner.

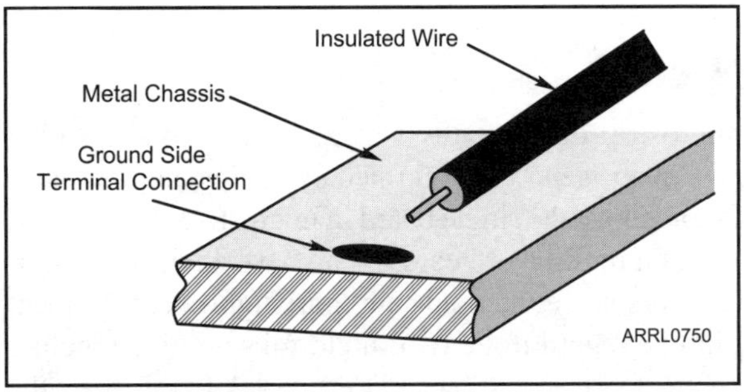

Figure 4.4 – An insulated wire positioned tightly against a chassis or other metal surface can serve as a transmission line similar to a microstrip line.

As with microstrip, while the end-to-end transmission is predictable and efficient, there can easily be unplanned and undesired coupling to other circuits within the chassis since the fields are not completely contained close to the wire. If there are other sensitive circuits close by, a piece of miniature coax can often fit in the same place and, if properly terminated, can reduce the coupling.

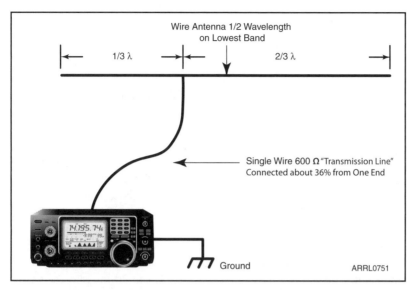

Figure 4.5 – The Windom antenna, c. 1929, used a single wire as a transmission line. There is controversy even today as to whether it really acts as a transmission line, or as a top-loaded vertical monopole.

SINGLE WIRE ISOLATED FROM GROUND

A popular multiband amateur antenna before World War II and continuing in popularity in different versions today is the Windom, as shown in **Figure 4.5**. It was described in detail in a 1929 *QST* article by L. G. Windom, based on work by others, and still carries his name.[1] The idea was that an antenna a half wavelength long on 80 meters could also be used on the other three HF bands of the time — 40, 20, and 10 meters.

However, there was a wide variation in the impedance at the center of such an antenna as the bands were changed. The solution was to feed it off center, at approximately one third of its length. The antenna would then provide a reasonable match to the line on all four bands.

The current in the "transmission line" has no direct return path to cancel its radiation, so it will radiate as well as (or perhaps better) than the antenna, depending on the band. In the typical configuration, this makes for a combination of vertically polarized radiation from the feed line and horizontally polarized radiation from the antenna. This may or may not be useful, depending on where the signal's desired destination is, but the RF at the bottom of the feed is almost never a good idea for the radio equipment or operators.

Recent Windom Implementations

More recently, Windom-type antennas have substituted high-impedance transmission lines for the single wire line. Some use a transformer at the antenna connection point to transform the impedance to 50 Ω for connection to coax. In both cases the lack of a balanced connection point at the antenna results in common mode current on the transmission line that will radiate. The coax-fed variety uses a choke at the bottom of the feed line to keep the current from the station. In both cases, the vertical section will radiate — commercial versions proclaim this as a selling point and have many happy users. One thing to keep in mind is that in Windom's day, we had four harmonically related bands. We still have those, but we've added five more, making an "all-band" antenna design even more of a challenge now than in 1929.

So while such a single wire may serve in some ways as a low-cost transmission line, and may have met the needs of early amateurs, it is not in the same category as lines in current use. It should thus be used with caution, if at all.

[1]L. Windom, W8GZ/W8ZG, "Notes on Ethereal Adornments — Practical Design Data for the Single-Wire-Fed Hertz Antenna," *QST*, Sep 1929, pp 19-23, 84.

Chapter 5
Let's Examine Balanced Transmission Line

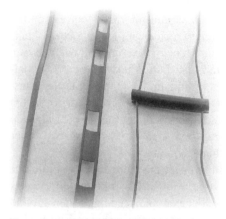

Three examples of balanced transmission line. From left, 300 Ω TV type twinlead, nominal 450 Ω "window" line and 600 Ω open-wire line.

In an unbalanced line of the type we have been discussing, the signal is applied between a conductor and a common conductor, often a shield or a ground plate. The voltage on the *hot* conductor is specified with respect to the common conductor. In a balanced line, there are two conductors, both at a potential with respect to ground. If properly balanced, both signals are of the same magnitude, but of opposite phase, so the total line voltage is twice the voltage of either with respect to ground.

This may be easier to grasp for those familiar with standard US household wiring. The transformer on the pole, which serves perhaps five houses, takes the multi kV distribution voltage and transforms it to 240 V ac. The secondary of the transformer has a center tap that is the *common* connection. The balanced pair and common are connected to the house, usually via a twisted wire cable. At the house end, the cable goes through a usage meter for billing purposes, then to the household distribution and circuit breaker or fuse box. The common connection is tied to the box frame, which is also tied to a ground connection at that point. Consider **Figure 5.1**, in which we portray the usual connectivity from the line to the primary ac distribution panel.

The two outer conductors are used to provide a balanced (with respect to common) 240 V feed for such heavy appliances as ovens, electric clothes dryers, or electric heating systems. Either outer conductor, fed with the common circuit, can be used to provide a 120 V unbalanced source for standard

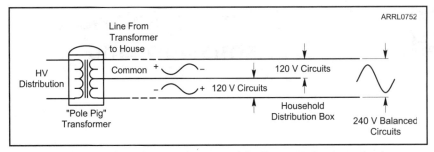

Figure 5.1 – Typical household ac power connectivity illustrates the concepts of balanced and unbalanced transmission lines. The 120 V circuits are unbalanced with a common return, while the 240 V circuits and their transmission system are balanced with respect to the common.

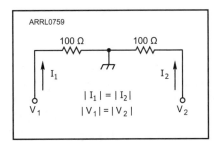

Figure 5.2 – Perfectly balanced load, where both the current and voltage on each side will be in balance.

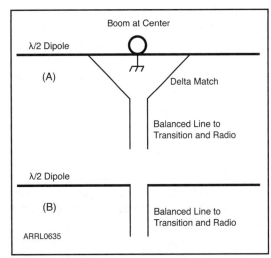

Figure 5.3 – Two examples of inherently balanced antennas, one (delta matched dipole) with a central ground (A) and one (split dipole) without (B).

wall outlets with the common serving as a return. This is exactly analogous to radio transmission lines. Note that while each 120 V circuit returns its current on the common wire, if the two sides are feeding equal loads there will be no net current on the common wire between the transformer and distribution box. If there is unbalance, the current returning on the common to the transformer is referred to as *common mode* current, a concept we will discuss later. This is in contrast to the balanced *differential mode* current that flows on the two outer wires — that's where we want it.

In terms of the more directly applicable radio uses, we can illustrate the differences as well. For example, feeding the non-ground end of two 100 Ω resistors (see **Figure 5.2**) that have the other end grounded will result in a balanced 200 Ω system. In such a case, the voltage on each side will be the same magnitude, but 180° out of phase. The magnitude of the currents on each side will also be the same, since the voltages and resistances are equal.

Note that the system would be balanced even without the ground connection. No current flows in the ground lead of the perfectly balanced system, so it could be removed without changing the operational properties. **Figure 5.3** shows two examples of inherently balanced antennas, one with a central ground

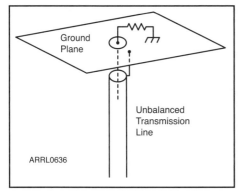

Figure 5.4 – Unbalanced resistive load. The signal is applied with reference to ground.

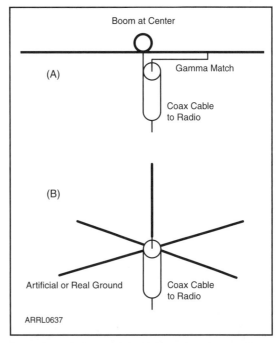

Figure 5.5 – Two examples of inherently unbalanced antennas, a gamma matched dipole at (A) and a vertical monopole at (B).

(A) and one without (B). The ground in the first is not actually necessary, but can be beneficial for lightning protection purposes. For both cases, some kind of transition is required to shift to an unbalanced system for connection to the radio. The types of transition will be the subject of a later chapter.

An unbalanced system, on the other hand, is fed with respect to ground. That is, one side of the load is at ground potential. **Figure 5.4** gives an example of an unbalanced resistive load, and **Figure 5.5** shows two antennas with inherently unbalanced feed points.

TYPES OF BALANCED LINE

Balanced transmission line is generally of relatively simple construction. Think two-wire lamp cord or speaker wire — both common examples of wire construction that could be used as balanced transmission line. Early receiving balanced line was typically a twisted pair of wires used to connect a dipole antenna to the commonly found balanced input of a receiving set of the time.

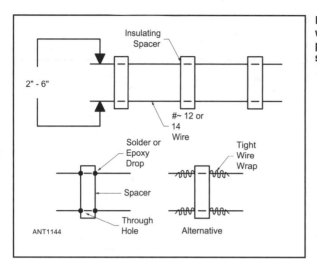

Figure 5.6 – Construction method of open-wire transmission line. Basically, just two parallel wires with a sufficient quantity of spacers to hold them apart.

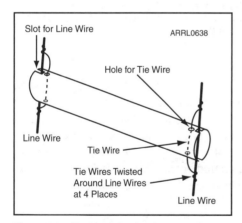

Figure 5.7 – Modern homemade open-wire line often uses PVC tubing as the insulating spacers. Easier fabrication is possible if the wires are in slots rather than being pulled through holes. The small tie wires run through holes to secure the line to the spacer.

For transmitting use, open-wire line was usually employed. It is still popular today for many applications and offers simplicity of construction, as shown in **Figures 5.6** and **5.7**. In the earliest days, the spacers were often made of wood that was boiled in paraffin to reduce the penetration of water. During World War II, and for some time after, porcelain spacers were made for the purpose. They are still ideal in many ways, but no longer in production.

Modern commercial balanced line is often made using a molded polyethylene web as both spacer and insulator. As shown in **Figure 5.8**, such line comes in several varieties. The twinlead types have continuous insulation, while the window line (also available in 300 Ω versions) has "windows" punched out periodically to make it almost equivalent to open-wire line. Twinlead at 300 Ω was originally developed as a transmission line for television reception, and was popular for connections to TV antennas until inexpensive low-loss coax overtook it in the marketplace

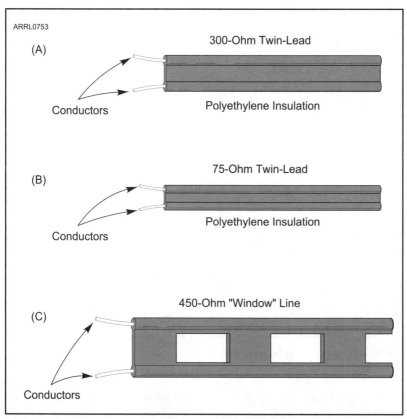

Figure 5.8 – Examples of balanced transmission line made with a molded polyethylene insulating webbing. Window line is made the same way, except that "windows" are punched out periodically to make the characteristics more similar to open-wire line.

in the 1970s. It is still available at electronic retailers, as are TV and FM antennas designed to use it. Open-wire line, constructed much like that shown in Figure 5.7, is available commercially as well.

There are a few varieties of shielded twinlead around as well. These have been in use for some time, first in military systems with various RG numbers, including RG-22 and RG-97 with a Z_0 of 95 Ω. Later, shielded 300 Ω twinlead was offered for TV antenna use in areas that would not be appropriate for usual balanced lines. Unfortunately, both have the advantages and disadvantages of unbalanced coax. They are easy to route through materials that would give problems to standard twinlead, but because of their construction, they also have loss characteristics similar to coax. If needed, a similar balanced transmission line can be fabricated from two coax cables,

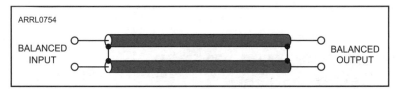

Figure 5.9 — Balanced transmission line made from two equal-length pieces of coax with shields tied together at each end. While it offers the installation flexibility of coax, it also has similar loss.

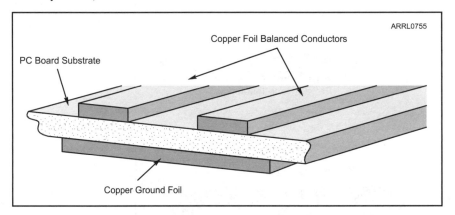

Figure 5.10 — Balanced microstrip line fabricated as part of a printed circuit.

as shown in **Figure 5.9**. The Z_0 is just twice the Z_0 of the individual cables. Note that if the shields are tied together at both ends, there will be no fields between the two cables, so they can even be run along different routes, as long as they are the same length. The loss will be the same as for coax.

Printed circuit board fabrication of stripline, or microstrip line, is also adaptable to balanced systems. The balanced microstrip line shown in **Figure 5.10** utilizes a balanced pair of conductors on the opposite side of the substrate from a ground foil to help contain the fields; however, a balanced pair on single-sided board is also possible.

THE BENEFITS OF BALANCED LINE

Balanced line has two primary benefits in comparison to coaxial cable. Most balanced lines, especially those with large portions of air dielectric, such as window line or ladder line, have a significantly lower loss than most coaxial cables.

Balanced line is often very conveniently connected to balanced antennas and is most often encountered in that application. It is also frequently found in very long runs from unbalanced antennas in which the losses (or cost) of coax would be prohibitive. In that case, it may be worth the effort to transition from unbalanced to balanced near the antenna and then back to unbalanced at the radio end.

Most balanced line is relatively inexpensive when compared to coax — often an important consideration. Excellent homemade open-wire line can be fabricated at low cost from two rolls of wire (stranded works very well, particularly if subject to flexing) and insulators made from inexpensive household PVC tubing.

THE DOWNSIDES OF BALANCED LINE

As might be expected, there are downsides to balanced line. While coax cable, with its fields contained within the shield, can be rolled, buried (if it has a direct burial rating), or installed within or near pipes without impacting performance, this is not true of balanced line. With balanced line, the fields conveying the signal down the line exist in a region around the two lines several times the distance between them. This means that balanced line can't be placed on the ground, run in metal ducts, run through lossy material, or be rolled up without causing additional loss or change in characteristics.

FIELDS BETWEEN CONDUCTORS OF BALANCED LINE

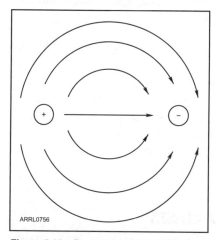

Figure 5.11 – Representation of the electric field between the conductors of balanced transmission line. The magnetic fields that surround each conductor are equal and of opposite sense if the currents are balanced.

The fields between the conductors of balanced lines are not constrained within a boundary, as is the case with coax. The strongest fields are in the space directly between the conductors, but the total fields between are similar to that shown in **Figure 5.11**. Outside the region between the conductors, the fields will tend to cancel in the plane at the center between and perpendicular to the wire axis. Coupling to other circuits will be strongest near either wire until the distance is large compared to the spacing. The coupling will be minimal for most purposes at a distance of two to three times the conductor spacing.

In the heyday of twinlead for TV use, tubular twinlead was popular for critical applications. This provided an elliptical cross-section of air space in the region between the conductors in an attempt to minimize losses in the region in which the fields are strongest.

While fields at a distance from balanced line are small, nearby fields can cause interference to other systems, particularly those interconnected by wiring.

In addition, balanced line run too close to computers and other potential RFI-generating systems can pick up interfering signals during reception.

ELECTRICAL CHARACTERISTICS OF BALANCED LINE

As with coax, a lumped circuit model of balanced transmission line can be generated based on the per unit electrical characteristics. A schematic diagram of the result is shown in **Figure 5.12**. According to Kraus, the per unit inductance of a balanced pair of similar wires removed from ground and with an air dielectric is:

$$L = 0.92 \times \log^{10} (D/r) \, \mu H/m, \text{ if } D \gg r$$

Where D is the center-center spacing, and r is the radius of each conductor in the same units.[1]

The characteristic impedance, under the same conditions, is:

$$Z_0 = 276 \times \log^{10} (D/r) \, \Omega$$

Since Z_0 also equals $\sqrt{(L/C)}$, we can easily find the capacitance as $C = L/Z_0^2$.

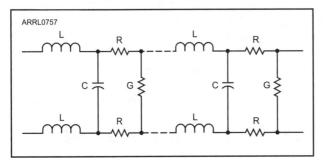

Figure 5.12 – Lumped element model of a balanced transmission line.

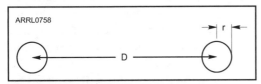

Figure 5.13 – Dimensions used to calculate transmission line parameters.

The series resistance is just the dc resistance of the wire at low frequencies, increasing at higher frequencies due to skin effect forcing current to the outer region of the conductors with corresponding reduction in effective surface area of the conductors.

The shunt conductance is the result of the losses in the dielectric. This is usually minimal in air dielectric line.

MATCHED LINE LOSS OF BALANCED LINES

The loss of matched balanced transmission line is provided in Table 3.1 in Chapter 3. It is worth making a note of the difference between the matched loss of coaxial cables shown in comparison to the losses of window line. Looking at the loss per 100 feet column of coax cable at 10 MHz, we note that most are in the range of 0.5 to 1.5 dB/100 feet. There are very large and very small cables outside this range, but typical matched loss would be around 1 dB/100 feet. The balanced lines shown are in the range of 0.06 to 0.3 dB/100 feet, with window and open-wire line all in the 0.06 to 0.08 dB/100 foot range. At the 100 MHz point, the difference is perhaps even more striking. Flexible (non-hardline) coax, in the usual sizes, runs from around 1 dB to more than 5 dB/100 feet, while the window lines are all 0.3 dB/100 feet or less.

MISMATCHED LINE LOSS IN BALANCED LINES

The low matched loss results in dramatically lower total line loss for window line in comparison to the usual coax. Let's compare two lines with an SWR of 10:1. If we had 100 feet of coax with a 5 db loss (see Figure 3.8 in Chapter 3), we would add an additional 4 dB of loss due to the mismatch for a total of 9 dB. For window line with a matched loss of 0.3 dB, we would add only 0.95 dB, for a total of about 1.3 dB.

If we put 100 W of transmit power into the coax, at the far end we will have 12.6 W available to radiate from our antenna — a loss of 87.4 W in the transmission line as heat. With the window line, the antenna will get 74.1 W, a loss of 35.9 W in the transmission line. While 1 dB is usually just perceivable, 9 db can make a big difference.

This is the big strength of balanced line, particularly window and open-wire lines at HF where the matched loss is dramatically small. With window line, antenna systems can be designed to operate with high SWR, making it easy to cover multiple bands, for example. Coax, on the other hand, is generally limited to uses with an SWR of no more than 2 or 3:1 to avoid excessive loss. It is particularly noteworthy that the best window line generally costs less than even inexpensive coax.

POWER HANDLING CAPABILITY OF BALANCED LINE

Table 3.1 does not provide any information about the power rating of the various balanced lines. While the dielectric strength of polyethylene will be well into the megavolt per inch range, the line breakdown will likely be limited by the dielectric strength of air at most interfaces, about 76 kV/inch. The wire will also fuse at some current, around 82 A for #18 AWG copper wire.

Keep in mind that both the maximum voltage and current on the line will go up with the square root of the SWR, compared to what they would be for a matched line. As an example, consider a 450 Ω window line at 1500 W, the US amateur power limit. The voltage if matched would be:

$V = \sqrt{(Z_0 \times P)} = \sqrt{(450 \times 1500)} = $ 822 V, with a 100:1 SWR, a maximum of 8222 V at some locations along the line.

The line current would be:

$I = \sqrt{(P/Z_0)} = \sqrt{(1500/450)} = 3.3$ A, with a 100:1 SWR, a maximum of 33 A at some locations along the line.

Neither of these quantities would be a problem for the physical characteristics of the line; however, the components to which they are connected in terminal equipment could be a very different story. Note that the maximum of one parameter in mismatched line will occur at the location along the line of the minimum of the other, with the maximums separated by ½ wavelength. Thus the maximums need not be encountered at all, and the line length can be designed to avoid maximums at sensitive locations, but only if the system will be used at a particular frequency. Multiband systems are likely to hit some maximum on some band.

[1]J. Kraus, W8JK (SK), *Electromagnetics*, McGraw-Hill Book Company, New York, 1953.

Chapter 6
Transmission Line Connectors

The popular "UHF" connector series is synonymous with "coax connector" for many, yet it is not always appropriate.

The most usual application of transmission lines is interconnecting different subsystems such as radios to antennas, interconnecting video systems or any functions that need to be interconnected in a well-defined way. A transmission line only becomes useful if the ends are properly prepared to interconnect with equipment at each end.

Generally, purchased equipment will include one side of the connector that the designer feels is appropriate for the application as part of the equipment. It is the responsibility of the systems integrator or installer to research the connector type and make sure the transmission line is terminated in a connector that will mate with the one on the equipment at each end to allow the connections to be made. Note that with different equipment types, the connector at each end of the cable may be different.

If you are designing the entire system, you have the flexibility to select the connector types as part of your design. It is important to select connectors that do not degrade system performance. This chapter will discuss the various connector types and their properties to help in that determination.

There are many types of connectors especially designed for termination of coaxial cables. These cables first came into widespread use in radio and video (radar screen data, for example) systems around the time of World War II due to the use of higher frequencies for communications and navigation systems for the military. In subsequent years there have been many new types developed for special purposes, some of which have come into general use.

"UHF" SERIES COAXIAL CONNECTORS

The UHF connector series was defined at a time when the "ultra high frequencies" meant anything above 30 MHz. Since we now consider VHF the range between 30 and 300 MHz and UHF from 300 to 3000 MHz, the designation is no longer appropriate, hence my use of quotation marks in the heading.

The first use of the UHF connector series was for broadband video systems such as radar signals. It allowed a termination of RG-8 sized coaxial cable into a continuously shielded interconnection across an interface. This was a significant improvement compared to the open screw terminals, binding posts, or Fahnestock clips commonly used in the prewar era.

The Basic UHF Connector Series

The connector consists of a center pin for the center conductor that plugs into a spring tensioned socket the size of a "banana" jack, surrounded by a screw-on backshell that provides the connection for the shield and holds the connection in place. **Figure 6.1** shows the basic elements of the system. The basic cable-terminating plug is still known by its World War II-era nomenclature, PL-259. The matching flange-type socket, nomenclature SO-239 (note difference in center digit), is normally used on the equipment side and mounts to a chassis or panel with four machine screws or other fasteners at the corners of the flange, which also serves as the shield termination.

Figure 6.1 – The basic elements of the UHF coaxial connector series. A PL-259 UHF plug is on the right, with a mating SO-239 socket on the left.

The original PL-259 design called for the center conductor to be soldered to the inside of the hollow pin while the outer conductor or shield is soldered inside the outer shell through the holes provided. The inside of the outer shell of a PL-259 includes course threads that can help secure the cable jacket if tightly screwed in before soldering. **Figure 6.2** shows a typical application of a UHF connector pair.

Figure 6.2 – A PL-259 plug attaches a cable to an SO-239 jack on a piece of equipment in the author's station.

Additional UHF Connector Types

Over the years there have been additional types of UHF connectors added to the series, some officially and some designed to meet particular needs.

Figure 6.3 – Additional UHF series connector types. At the top is a PL-258 coax splice with an extended version below. Next is a UHF male-female elbow adapter that allows a change in cable direction — handy with heavy cables. At the bottom a double-male adapter.

Figure 6.4 – A PL-259 with UG-175 reducer and an installed RG-58 cable.

Double Female Adapter: Probably the first was a double female adapter that allowed the interconnection of two male connectors, effectively splicing two cables together. This has a military designation of PL-258 and is shown at the top of **Figure 6.3**. Because of their shape, they are often referred to as *barrel* connectors. An unofficial expansion of the PL-258 concept is a somewhat similar double female connector that is made longer with a continuously threaded outer surface. These are usually provided with matching nuts (see one below the standard PL-258 in Figure 6.3) designed to secure them going through panels, partitions, or even walls. I have seen them up to 1 foot long.

Elbow Connectors: A male-female elbow connector (below the PL-258s in Figure 6.3) is often useful for changing a cable's angle of entry into equipment. This will allow a cable to hang straight down behind equipment without requiring a tight radius bend.

Double Male Adapter: Not surprisingly, double male adapters (shown at the bottom of Figure 6.3) are occasionally encountered as well, although the need to interconnect two pieces of equipment at close range is less common.

Reducing Adapters: The sizing of both the inside of the pin and the inside of the shell is just right for the 0.405 inch outside diameter of RG-8 or RG-11 and their inner conductors, or for their successor types of cables of similar size.

In many applications, it is desirable to use the smaller, more flexible RG-58 or RG-59 cables, especially for short patch cables. The shields of these cables are too small to be properly soldered to the inside of the PL-259 plug shell.

UHF connector adapters are available that screw into the back of the shell and provide a proper fit to the outer jacket of the smaller cable. The shield is then folded back over the adapter to appear at the plug solder holes. While the center conductor is also smaller, generally the pin can be filled with solder to complete the connection. The adapters, sometimes called reducers, are available in a number of sizes with the most common being the UG-175 for RG-58 size cables and UG-176 for RG-59 size cables including RG-8X. A UG-176 adapter with a PL-259 and an installed cable end are shown in **Figure 6.4**.

T Connectors: There are also triple-connector T connectors, sometimes with all three ports female, sometimes with the center port a male plug. These are useful for bridging cable connections, but consider the implications of parallel impedances to the system SWR. Another variation that is frequently encountered is a jack with a threaded body that mounts through a panel hole much like half of the extended double female.

Jack Hood: In order to address the unshielded nature of the standard SO-239 panel jack, a special hood is also available that provides a solder connection to the shield and a shielded surrounding of the center conductor. While this can improve isolation, it exacerbates the Z_0 issue because of the changing outer conductor diameter of the hood's conical shape.

Jacks Without Flange: Jacks are available that mount into a single ⅝ inch diameter hole. The threads for the backshell are extended to the rear of the connector, and a shoulder is located on the panel side. It looks superficially like a PL-258, but has a center solder terminal.

Limitations of UHF Series Connectors

That the UHF connector is as popular as it is may be a testament to the power of being early on the scene. While it can be useful, it does have many limitations and weaknesses, none of which make it unsuitable, but all of them need to be taken into account. The major limitations, in my view, are listed below.

♦ The characteristic impedance of the connector itself is not the same as that of the usual coaxial cables with which it is used. The connector pair Z_0 is typically 30-40 Ω for perhaps an inch of length. While there actually are 36 Ω cables for which this would be a plus, they are rarely encountered. This results in an "impedance bump" that can cause a change in the impedance at the input to the system, depending on the electrical length.

The significance of this limitation will depend on how critical the cable impedance match, or SWR, is to the system operation. It is also true that it will be much more significant at frequencies with wavelengths above about 10 times that length or 10 inches. This corresponds to just above 1 GHz, within the UHF range. Many elect to use other connector series above 6 or 2 meters.

♦ UHF connectors are not waterproof. This is not usually a problem with their use indoors, but can be a serious problem for outdoor use, such as on antennas. Water not only causes problems with the integrity of the connector itself, but can also result in water migration into the cable. While water in a cable is never a good thing, it gets worse over time, often resulting in corrosion and contamination of the dielectric. This can result in additional cable loss over the years.

Fortunately, the connectors can be waterproofed using various tape-like products intended for the purpose. This is an extra step and one that impedes the connection and disconnection of outdoor equipment.

♦ The shield connection primarily occurs at the perpendicular surface between the two connectors, augmented by the backshell itself. The integrity of both aspects of this connection is dependent on the tightness of the backshell. In mobile systems, or others subject to vibration, this can result in loose, high resistance, or even disconnected shields over time. Such conditions result in excessive loss, noise, undesired coupling, and functionally the loss of all benefit of coax in the first place. Many recommend the use of pliers for an extra bit of torque, rather than just hand-tightening the backshells.

♦ Availability of low-quality products. While this is not a fault of the connector design, it is a fact that the UHF series is relatively easy to duplicate. It has also been around long enough so that there are many substandard UHF connectors in the marketplace, which may present the following issues:

- Inner conductor plug pins that are too narrow to accommodate RG-8 center conductors.
- Inner conductor plug pins or connector sockets that fall out of the body.
- Plug insulation that melts while soldering.
- Poor-quality metal parts, including insufficient or improper material plating.
- Jack spring fingers that lose tension after a few insertions.

The solution is simple — only buy products from manufacturers known to make high-quality products. The one that I know is always of high quality is Amphenol, although I'm sure there are others. All major Amateur Radio dealers offer Amphenol UHF connectors, and the prices are reasonable. I recommend avoiding electronic retail store house brands, as well as any connectors found at flea markets that don't show a known high-quality brand.

TYPE N COAXIAL CONNECTORS

The Type N connectors are of a similar size to their UHF counterpart, but these connectors are seriously designed for UHF operation. They were designed in the 1940s by Paul Neill of Bell Laboratories to overcome the limitations associated with the UHF type, particularly with respect to higher frequency operation. While originally specified to work well up to 1 GHz, they are routinely used at frequencies more than 10 times that high. The nomenclature is in appreciation of Neill's name.

The Basic Type N Connector Series

Only a very close look at a connected pair will indicate that it's not a UHF pair, however, once they are apart, the difference is clear. The biggest difference is that the shield interconnection is primarily made via a spring

Figure 6.5 – A Type N connector pair — a good choice as an alternate in any application for which a UHF type would be used, but especially beneficial in the current UHF range or in wet conditions.

fingered sleeve surrounding the male pin that fits snugly into a matching cylinder in the female jack. This means the shield will be connected even if the backshell is loose.

A Type N connector pair is shown in **Figure 6.5**. The concentric shield connection on each side is evident in the photo. The size of the shield connection, the air space in between, and the mated pin/socket combination make for a constant-impedance, low-loss match.

There are two versions available; one with center pin and socket diameters that have a Z_0 of 50 Ω, generally used for RF work, and another at 75 Ω intended for video applications. The constant impedance avoids the impedance bump associated with the UHF series — the major frequency limiting factor of the UHF series. To put frosting on the cake, properly assembled Type N connectors are also waterproof.

As discussed detail in the *Coaxial Connector Assembly* chapter, the traditional Type N connector is assembled in a different manner than the soldered UHF type. A clamping arrangement is used to assemble the connector with the shield compressed between a washer and the shoulder of both sides of the connector. This avoids the melting cable dielectric insulation problem encountered with the soldering of shields to some kinds of connectors, but at a cost of requiring more careful measurement of trimming dimensions to make all the pieces come together in the right place. The only soldering required is of the low mass center conductor to its pin.

Other Varieties of Type N Connectors

Virtually all the different connector types described in the section under UHF connectors are also available for Type N connectors. A selection of Type N connectors is shown in **Figure 6.6**. In addition, the Type N series also offers a cable jack that looks like a cable plug, but is of opposite gender. This is significant because it allows the fabrication of extension cables that can be spliced together without the need for a barrel, or double female adapter. This is beneficial because each connector interface contributes to system loss and reduction of system reliability.

Figure 6.6 –Type N series connector variations. At the top, a Type N cable socket, below double female, double male and right angle adapters. Not shown — T adapters. All of the varieties of UHF connectors are also available in Type N.

Limitations of Type N Connectors

Here we have a much shorter section than for UHF connectors. As with all connector types, there is a power limitation associated with the connectors, basically related to its physical dimensions.

Power Limits

The peak power is limited by the voltage breakdown across the air gap between the center pin and the shield connection, with a safe recommendation of at least 1000 V_{RMS} corresponding to a power level of 50 kW, while the average power is based on the heating of the conductors, limited by the current in the center pin. Skin effect concentrates the current in thinner and thinner outer regions of the pin as frequency increases, resulting in a typical average power rating around 5 kW at 20 MHz, decreasing with frequency to around 500 W at 2 GHz. Check the connector manufacturer's (as well as the cable manufacturer's) specifications if you expect to approach these limits.

Impedance Variation

This is a more insidious issue. The two different versions of the Type N connector have exactly the same size backshell and shield connection arrangement. The difference between them is strictly with the diameter of the center conductor, the 75 Ω version being thinner to result in a higher Z_0, as you would expect.

This means that a 75 Ω plug will mechanically fit into a 50 Ω Type N socket without difficulty. Unfortunately, it will not make a good contact. I am aware of one case in which the assembly alignment of the two was so perfect that it didn't make contact at all, leading to a lot of head-scratching.

The problem in the other direction is actually worse, in a way. Inserting a 50 Ω Type N plug into a 75 Ω socket will cause the center female socket to open further than it should. It will likely work in this situation, however, if a 75 Ω plug is ever inserted, it will no longer make good contact.

This problem is exacerbated by the fact that the connector types have no distinguishing marks to identify them. The only reason that this doesn't often come up is that the two varieties are generally used in different domains — 50 Ω in RF work — 75 Ω in video systems. Should you find yourself in a mixed environment — over-the-air television transmission comes to mind — I suggest adopting a color-coded or tagging standard for all Type N connectors so that this is less likely to happen. Better yet, select a different connector type for one or the other application.

Assembly Process

Some would consider it an advantage for UHF connectors that most experienced technicians can strip the outer and inner insulation and terminate a plug by eyeballing the dimensions, while the Type N requires careful measurement to assemble properly. Especially with full size cable, the Type N has an advantage in not needing soldering of the shield connection, a time-consuming and failure-prone part of the operation. I would call it about even — but others will have their own opinion, especially if in the field without a rule or stripping guide.

BNC COAXIAL CONNECTORS

The BNC connector is perhaps the second most popular RF coaxial connector type in Amateur Radio circles. It was developed in 1951 and is related to the Type N, but has a smaller outer structure and a snap-on "bayonet" mount rather than the screw-on backshell of the Type N and UHF (see **Figure 6.7**). Its size is a perfect fit for the smaller (RG-58/59/8X) cables, although versions are available for smaller (e.g. RG-174) and larger (e.g. RG-8) sized cables.

Figure 6.7 – The basic elements of the BNC coaxial connector series. On the left is a BNC flange mounting jack. On the right is a BNC through-hole mount jack and in the middle, a plug.

The connector was invented by Paul Neill of Bell Laboratories (of Type N fame) and Carl Concelman (inverter of the Type C connector, below) of Amphenol. BNC stands for Bayonet — Neill — Concelman. The flexibility of the small cables and the easy to attach mechanism of the bayonet backshell, make BNC cables ideal for patch panel use.

When originally designed, the connector was just made to match 50 Ω systems — in fact the inner pin and shield connecting portion is the same size as that of the Type N (in fact, a Type N plug can be inserted into a BNC socket). Later a 75 Ω version was added, although the 50 Ω is much more frequently encountered. So now there are both 50 and 75 Ω versions of the BNC connector. However, this time they were cleverly designed to be interchangeable without destruction or connectivity issues.

The ratings of the BNC are quite similar to those of the Type N, since it has the same electrical dimensions, although its shielding is

Figure 6.8 – Additional BNC series connector types. All the usual suspects are available.

somewhat less effective so its upper frequency limit is around 2 GHz. In fact, the voltage rating may be higher because an insulating sleeve surrounds the center pin, making for a longer air path between conductors. All of the usual variations and special fittings for other connector types are available for the BNC (see **Figure 6.8**)

RCA CONNECTORS

The RCA connector, sometimes called a *phono* (for phonograph, an early music playback system) connector is probably more frequently encountered than any other type in Amateur Radio equipment because it is used for many other purposes besides terminating coaxial cables.

This connector was initially developed as an audio connector by the Radio Corporation of America to connect phonographs to amplifiers. Its structure is inherently coaxial with some jack configurations designed to ground shield connections directly to panels, so it has appeared as an RF connector as well. It is most frequently encountered as a receive antenna connector, especially in lower-priced equipment. As a transmit RF connector, it generally is found at power levels of 100 W or less, also in lower-priced gear.

A notable exception is the series of high-end Collins Radio SSB transmitters, receivers, and transceivers introduced in the late 1950s. These used RCA connectors for audio and RF interconnections. This highlights one aspect of this connector series — they are made in many forms. The RCA jacks on the Collins gear were of very high quality, using ceramic dielectric insulators. Some have been described as "RF rated" — a noble thought, but by no means a standard designation. Most usually encountered RCA connectors are not as well fabricated, but they can do the job if carefully selected and properly installed. A selection of different types is shown in **Figure 6.9**.

Figure 6.9 – Examples of RCA connectors. The plug at the bottom has a shielded backshell, more suited for RF use.

Unlike any of the previously described coaxial connectors, the RCA pair has no mechanical locking mechanism to retain the connections if subject to any pull on the cable. Since the BNC is similar in size, it may be a good alternative if that is likely to be an issue.

Since RCA connectors are so ubiquitous, it is worth mentioning that they are not, in my opinion, a very good choice for power interconnections. I have what is otherwise high-quality equipment that has RCA connectors as power accessory sockets and companion equipment with a panel jack designed to accept 12 V dc from an RCA plug. Of course if the equipment is on while making this connection, as soon as the pin strays to the outer conductor on insertion, it blows a fuse deep in the source equipment.

OTHER COAXIAL CONNECTORS

While the connectors described in the earlier sections are those most frequently encountered as coaxial cable terminations, there are many other types that will be encountered from time to time. I think it's worth a bit of ink to let you know something about them.

SMA Coaxial Connectors

SMA connectors — short for *sub-miniature type A* — are found on a significant number of handheld transceivers as antenna connections (see **Figure 6.10**). These small threaded coax connectors are designed for operation well into the microwave region — typically to 13 GHz, but some varieties go much higher. Outside of amateur handheld antennas, they are mostly found used as interconnections between chassis of microwave equipment, often on the ends of miniature semi-rigid coax.

Figure 6.10 – A handheld transceiver BNC connected antenna compared to an SMA connector.

Their electrical properties are quite suitable for amateur handheld antennas, however, they are typically rated for 500 to 600 connect-disconnect cycles, so care should be taken with their mechanical properties. The alternate connector in this application is the BNC, which is my preference for a number of reasons, especially durability. An SMA handheld antenna is shown in Figure 6.10.

Type F Coaxial Connectors

Anyone who has ever hooked up a modern television set has encountered the Type F connector. This is a great mass-market connector — as inexpensive as possible for millions of users. I remember a time (more than 30 years ago when I was working on early cable-based digital networks using cable TV components) when the plugs cost less than $0.05/unit — orders of magnitude less than some we've discussed. The reason was that they made large quantities and the center conductor pin is just the inner conductor wire of the coax (see **Figure 6.11**).

With a center pin that is the center conductor of the coax, the assembly is trivial. The cable is stripped — fancy tools that do it in one step are available — the shield is crimped to the outer conductor shell of the connector (while soldering is possible with some varieties, anyone using these

Figure 6.11 – Examples of Type F connectors as used in cable TV installations.

frequently uses a crimp tool), and you're done. The disadvantage is that the inner conductor must be solid wire and that it has no corrosion resistance. Special waterproof versions are available for outdoor use.

Type F connectors are not often encountered in Amateur Radio, with one notable and worthwhile exception. TV-type coax cable, often described as RG-6 type but made with aluminum foil shields, is actually pretty good coax. Modern cable TV systems operate well into the UHF range, so they are designed for low loss. They are often available at low cost, or even prewired within a house begging to be used for amateur antenna access. The problem is that most connectors of the type used in Amateur Radio can't deal with connections to the aluminum foil shield. Often the best approach is to either use the F connectors on the cable, or crimp some on and then use between series adapters (see below) to transition to the connector that will mate with your termination.

TNC Coaxial Connector

The TNC (Threaded — Neill — Concelman) is essentially a BNC with a threaded rather than a bayonet backshell. It thus is essentially a smaller Type N connector, but with the inner connecting surfaces the same size. It extends the 2 GHz frequency limit of the BNC up to 10 GHz or higher, and takes less panel space. It is a natural for heavy duty requirements with smaller cables.

Type C Coaxial Connector

This connector, honoring just Concelman, is essentially a TNC for larger cables. It is about the size of a Type N, but with a bayonet backshell and larger diameter inner components. It would seem to be a natural for patch panels for those running high power, but it has never caught on in the amateur community, that I'm aware of. In fact I hardly see them anywhere, which is too bad.

Motorola Connectors

These are the coaxial connectors designed and almost exclusively used to connect automotive antennas to automotive broadcast radios (see

Figure 6.12 – A Motorola jack. These are commonly used in automotive broadcast receivers as antenna connectors.

Figure 6.12). They became popular for amateur use in the 1950s, at which time most HF voice used full carrier amplitude modulation. A number of companies offered tunable converters, designed for steering-column mounting, that translated amateur HF signals to the upper broadcast band.

The auto antenna, with its Motorola connector, plugged into the converter and a patch cable went from the converter to the antenna jack of broadcast radio. Usually, each had a Motorola connector. Apparently because of the converter connection, I have seen these connectors also used in home station VHF converters.

Type HN Coaxial Connector

The Type HN connector is similar in design and layout to the Type N, but is a bit larger. It has higher voltage breakdown and power ratings than the type N. It is a good match to the cables that are somewhat larger than RG-8 size and finds some amateur use in high power stations with long cable runs.

General Radio Connectors

The General Radio type 874 or GR connector is most frequently seen on test equipment made by that company, later GenRad through the 1970s. It is still popular in university physics and engineering laboratories because it is unique (to my knowledge) as a genderless coaxial connector. Both the

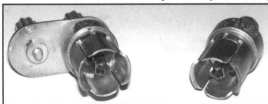

Figure 6.13 – The GR-874 genderless coaxial cable connector — popular for use with laboratory test equipment for many years.

inner and outer connection arrangements are made of four spring leaves in two pairs, with one pair of each separated further than the other (see **Figure 6.13**). In this way, by rotating 90 degrees, any two connectors can be mated resulting in a solid, matched connection without the need for gender converters (see **Figure 6.14**). The downside is that there is no locking arrangement, so it's not particularly suitable for long-term connections, but it's perfect for lab use.

Figure 6.14 – A pair of GR-874 connectors mated.

Between Series Adapters

Virtually every coax connector type of either gender is available in a short adapter with a connector of every other type. These adapters allow equipment with one type of coax connector to be attached to cables

Figure 6.15 – Representative example of coaxial between series adapters.

terminated with another type. A collection of representative examples is shown in **Figure 6.15**.

These are best used for temporary test or measurement applications, although some find use in long-term applications. Every additional part, especially connectors, reduces the overall system reliability by adding another potential failure point.

There are a few exceptions, in my opinion, to this rule. One is that some equipment supplied with UHF jacks makes use of lower-cost parts in which the fingers give out quickly. While the best solution is to replace them with higher-quality units, sometimes that is not feasible and will likely void any equipment warranty. An adapter with a UHF plug and Type N socket is perfect for this situation, especially if the cable can be equipped with a Type N plug. This works best if done before the spring contacts get limp.

BALANCED LINE INTERCONNECTIONS

Considering how much longer balanced transmission line has been in use than coaxial cable, you might think there would be at least as many choices of connector. As it happens, that's not the case, but a few standards have evolved over the years.

Screw Terminals — Most Amateur Radio equipment designed for balanced transmission line offered screw terminals, as did most television receivers until the shift to coaxial cable interconnection made TV twinlead almost obsolete.

Screw terminals have the advantage of being nearly universal, in that virtually any reasonably sized wire size can fit under the screw head and may stay in place and not short during tightening. Still, the connection is not a terribly reliable one.

Most amateur receivers actually had three screw terminals. Two were for a balanced antenna, the third was a chassis ground (see **Figure 6.16**). The chassis ground should be connected to the power system safety ground, since many receivers had two-wire ac plugs. If an unbalanced antenna were

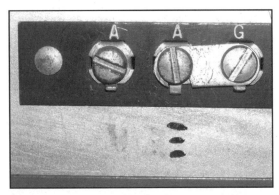

Figure 6.16 – Terminal strip on a 1950s communication receiver. There are two terminals for a balanced transmission line. If an unbalanced antenna is used, the ground terminal is jumpered to one of the antenna terminals as shown.

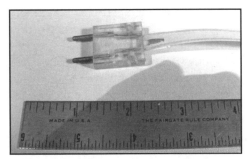

Figure 6.17 – A plug designed to terminate 300 Ω TV type twinlead.

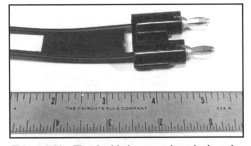

Figure 6.18 – The double banana plug, designed for low frequency balanced test equipment. It is a great fit with 450 Ω window line.

used, the ground should also be jumpered to one of the antenna connections, sometimes a link was provided for the purpose.

While wires under screw heads can be particularly unreliable, in my experience, a better solution is available in the use of crimped or soldered spade lugs under the screw terminals.

TV Twinlead Connectors

There actually was a type of connector especially designed for use with TV type 300 Ω twinlead (see **Figure 6.17**). For some reason, it never became popular for use with TV receivers, but was encountered in some peripheral equipment, such as UHF converters, before they were required to be included. The plug had two 0.093 inch soldered pins spaced at 0.486 inches. This is the same spacing as the pins of an FT-243 crystal holder, so would fit into a crystal socket, or two pins (with one in between) of an octal tube socket.

Double Banana Plug and Jack

A connector type that is not especially designed for the purpose is the double banana plug. A banana plug is a robust spring loaded plug of a size that happens to fit nicely into the center conductor of a UHF socket. A popular test equipment connectivity arrangement is two banana plugs within a common insulating housing spaced 0.75 inches (see **Figure 6.18**). This spacing is very convenient for use with most window-type transmission line.

Jacks are available in various forms, including panel mounting types and even combination plug and jack sets that also have screw terminals to clamp wire leads. Another handy arrangement is a kind of terminal that is sometimes called a "five-in-one." This mounts through a single hole into a panel and has a banana socket surrounded

Figure 6.19 – A five-in-one terminal accepts a banana plug, loose wire, or terminal lugs. If two are spaced at 0.75 inches, they will also accept the handy double banana plug.

by a nut arrangement that surrounds a threaded surface that includes a hole for wire or tip insertion. Thus it can accept a banana plug, a spade lug, a wire bent around the threads or a wire pushed through the perpendicular hole. If two are installed on ¾ inch centers, it will also accept the double banana plug, but can deal with loose wire ends, or lugs as well (see **Figure 6.19**).

Notes
[1]J. Kraus, W8JK (SK), *Electromagnetics*, McGraw-Hill Book Company, New York, 1953.

Chapter 7

Installing Coaxial Connectors on Cable

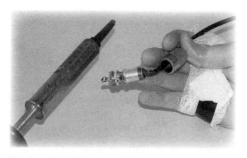

No coaxial cable run is any better than the installation of the connectors on its ends. Connector installation takes care and attention, but need not result in a trip to the emergency room.

The installation of transmission line connectors is looked on by some with dread, since it is both important and requires special care and moderate precision in order to do the job properly. Of course, any connector installation will be no better than the quality of the connector being used. Connector failures are surely the leading cause of problems with coax cables, although there are certainly other ways to ruin a coax run.

Standard or Crimp Connectors?

The "standard" coaxial cable designs are generally assembled by soldering. In the case of the PL-259 UHF plug, both the center conductor and shield are soldered, while standard Type N, BNC, and TNC connectors have a center pin that must be soldered and an integral tightened clamp for shield connection. The Type F television coax connector is a bit different. The center conductor of the coax is used directly as the center pin, while the shield is almost always crimped.

Alternative connector types are available in each series that use crimping for connection. In many cases, both the center conductor and shield are crimped, while some have a crimped shield and soldered inner conductor. I have seen no evidence that a crimped connector is inferior to a soldered one. With either type, poor assembly practices can yield poor results.

The issue with crimped connectors is that they require a crimping tool that is matched to the sizes of the crimped surfaces. Most quality crimping tools are relatively expensive and fit a limited series of cables and connectors. Some have dies that include multiple sizes of crimping holes, making

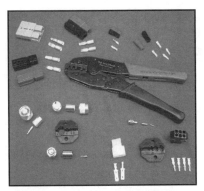

Figure 7.1 — The Andy-Crimp Pro™ tool, showing some of the available interchangeable dies along with a sampling of the different coax and power connectors that it can crimp with the right dies.

them more useful. The Andy-Crimp Pro™ tool — available from High Sierra Antennas (**www.hamcq.com**) or Quicksilver Radio (**www.qsradio.com**) — has available interchangeable dies that will also fit in other similar tools, such as the Powerpole crimper sold by West Mountain Radio (**www.westmountainradio.com**). In addition to the two die sets available for coax connections, there are sets available for the multiple sizes of popular Anderson Powerpole dc power connectors, and another set for Molex connectors and other crimp terminals (see **Figure 7.1**).

The crimping process is very effective if the proper tools and connectors that match the cable are all at hand. Some vendors have offered crimping arrangements for standard connectors, but these can result in fractures of the solid connector body. Crimping material needs to be pliable enough to be able to be formed in a press.

Another possibility is to make use of preassembled coaxial cables with the connectors already installed. As with anything else, the quality will vary depending on the capabilities of the supplier, often but not always indicated by the price. Suppliers with good reputations include ABR Industries (**www.abrind.com**), which provides cables with either UHF or N connectors, and DX Engineering (**www.dxengineering.com**), which provides cables with UHF connectors. These cable assemblies are available in different lengths and made from different types of cable, for a cost that is not much higher than the cost of the cable plus the connectors if you were to make them yourself. Both offer custom lengths and free shipping if a sufficient quantity is ordered.

The sections that follow will focus on installing standard coaxial connectors, with the exception of one representative sample of an Amphenol crimp-type UHF plug. If you decide to use crimp-type connectors and associated tools, the manufacturer should provide the cable and connector specific trimming dimensions. While the crimp diameters have moved toward industry standards, the details, especially of ferrule length, may result in different stripping lengths. Look at the website of LNL Distributors (**www.lnl.com/howto.htm**) to see some representative examples.

INSTALLING UHF SERIES COAXIAL CONNECTORS

UHF series connectors include cable-mounted plugs (the PL-259 type is the original pattern, although there have been others through the years) and chassis-mounted sockets (the SO-239 — note the difference in the center digit — being the standard type, although there are others here as well).

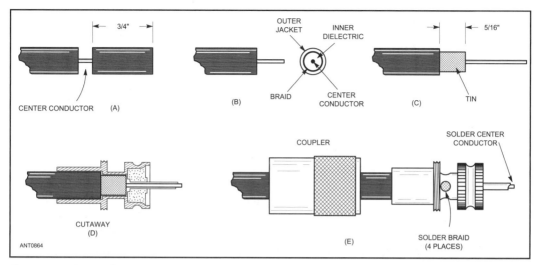

Figure 7.2 — The PL-259 or UHF plug is the most popular coax connector in Amateur Radio. Here the steps required for installation on RG-8/11/213 size (0.405-inch outer diameter) are shown. See the main text for a description of each step.

Most of the problems people have assembling these are with the plugs, so we'll start there.

UHF Plugs with Standard Cable

The standard PL-259 plug is designed to fit the common larger sizes of coax, such as RG-8, 11, 213 and RG-214, with an outer diameter of 0.405 inches. Note that this includes both 50 and 75 Ω types. Adapters are available for smaller cable types, discussed below. The installation of such coax into a PL-259 is illustrated in **Figure 7.2**.

In Step (A), the coax is stripped through the outer jacket, shield, and dielectric down to the center conductor starting ¾ of an inch back from the end, as shown. The hard part is cutting through all those layers and not also cutting some of the center conductor strands. This looks easier than it is with normal hand tools, but it is dramatically easier using a special tool. One by Ripley Brothers, the Cablematic UT-8000, works like a handheld pencil sharpener and makes a remarkably clean and straight cut.

By using a sharp knife, and going slow with a lot of patience, it is possible to work your way through the layers. Special tools are available that make this quite easy. It is also possible to use a tubing cutter with a sharp blade, at least to get through the outer jacket. Often, by the time you get all of that off, you will find that you have lost a strand or two of the center

Installing Coaxial Connectors on Cable   7-3

conductor. The purist would start over; you may decide to define your own quality control standard and decide to allow the loss of a strand, as it will not make much difference in the scope of things. (The perfectionist will certainly start over; it's always a good idea — one of many reasons to start with a somewhat longer cable than you think you might need.)

As shown in (B), remove the cut pieces together, if possible. Sometimes a bit of a tug with electricians' pliers is helpful. Once all of the dielectric is removed, carefully check, with a magnifier if needed, to make sure that there are no strands of the shield hanging about and that the center conductor strands aren't nicked.

In step (C), score the outer jacket $5/16$ of an inch back from the earlier cut without nicking the shield, and remove that section of the jacket. Sometimes a longitudinal cut from the score to the end makes it a bit easier to remove. Now *very lightly* tin the exposed shield. This is a very tricky step since:

◆ The shield has to be hot enough for the solder to flow into the wires without melting the dielectric. This is a particular challenge with foam dielectric type cables.

◆ The tinned shield needs to be almost the width of an untinned tightly woven shield in order to fit inside the barrel of the connector. This is a difficult step, since once the solder flows, it tends to wick into the shield. Recheck for any loose strands of shield that could short out the cable later. Some installers prefer to lightly tin the center conductor; this is not a bad idea, but also needs to be done sparingly to make sure the center conductor still fits within the pin. This can be checked from the pin side of the connector.

Now slide the coupling ring onto the cable, making sure that it is oriented in the proper direction. If doing both ends of the cable, this step can be delayed until the second end — just don't forget to do it. There is almost nothing less useful than a PL-259 without a coupling ring (or perhaps even worse, one remembered but installed backwards).

In step (D), the coax cable is inserted into the body of the connector. Again, check that the coupling ring hasn't fallen off. The connector should be installed with a clockwise turning motion in order that the threaded portion of the body screws onto the outer jacket. This can be a bit of an effort. Sometimes inserting the connector into the jack of a T connector will provide additional purchase, if the sharp points are engaged. Alternately, there are tools available, such as the DX Engineering UT-80P, that provide a good grip without chewing up the connector with pliers. The connector is on far enough if the center conductor is protruding from the pin and the shield is beyond the four solder holes. Recheck that the coupling ring is still there and oriented properly — this is your last chance for an easy correction.

Step (E) is the last part of the assembly process — soldering everything in place. First, support the cable and connector solidly, preferably in a way that doesn't draw off the heat from the connector body. Make sure that the cable weight is not pulling it downward, or else the heat from soldering will allow the dielectric to melt and the inner conductor to migrate through.

Use a heavy soldering iron of at least 150 W with a tip that will fit into the groove containing the solder holes. Since heat flows upward by convection, apply the iron to the bottom of the area and start flowing the solder into the top hole when the braid reaches melting temperature. Continue with the other holes, turning the cable carefully to reach the bottom hole. Let the cable and connector cool significantly before moving to the next step.

Check that the cable is not shorted by checking the resistance between the connector body and the inner conductor protruding from the pin. It should indicate infinite resistance, unless there is something providing connectivity at the far end. Now insert the connector in the vise with the open end of the pin pointing below the horizontal. Apply the soldering iron to the outside of the pin and gently apply solder to the junction of the center conductor and pin until it fills the void. When heated, the solder will flow between the inside of the pin and the center conductor by capillary action. If the pin were pointed the other way, it would flow down the outside of the pin and make it too thick it fit.

Let it cool and recheck the resistance. Trim the center conductor flush with the center pin and use a fine-tooth file to remove any solder from the outside of the pin. You're finally done — congratulations!

Another Opinion

Noted Amateur Radio contester Tim Duffy, K3LR, operator of one of the major multi-multi (multiple operator, multiple simultaneous transmitter) contest stations in the US, has another approach to assembling PL-259 plugs with the larger size cable, which he credits to William Maxson, N4AR. Tim has assembled an extreme antenna farm, visible from US Route 80 in Western Pennsylvania, and has terminated more coax cables than most people. Space considerations don't allow inclusion of all of his material, which can be seen on his website at **www.k3lr.com/engineering/pl259/**.

Tim first measures a connector against the RG-8 or other 0.405-inch cable and removes just the outer jacket with a sharp knife. He then pulls the braid back all the way by fanning it out (as shown in **Figure 7.3**) and

Figure 7.3 — After removing the outer jacket with a sharp knife, Tim carefully removes all but about ¼ inch of braid and fans out the remainder. [ROBERT BASTONE, WC3O]

wraps three to four turns of Scotch brand 88 black electrical tape around the dielectric up against the fanned-out braid (**Figure 7.4**). Using wire strippers, he removes the remaining dielectric from the center conductor (**Figure 7.5**). Then he installs the fully assembled PL-259 onto the center conductor and over the 88 tape. The back of the PL-259 body should rest on the fanned-out shield. He then solders the center conductor.

This done, he fans out the shield and cuts it to ¼-inch long and folds it over the back of the PL-259 (**Figure 7.6**). He solders the shield all the way around to the back of the PL-259 body. While the back is still hot, he wraps two turns of 88 electrical tape around the soldered shield to seal it all (**Figure 7.7**).

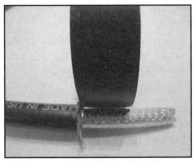

Figure 7.4 — Tim wraps three to four turns of electrical tape around the dielectric up against the fanned-out braid. [ROBERT BASTONE, WC3O]

Figure 7.5 — The dielectric around the center conductor is removed from the area in front of the tape. [ROBERT BASTONE, WC3O]

Figure 7.6 — Tim screws the connector onto the cable until it is fully seated, trims the shield, and solders the center conductor. [ROBERT BASTONE, WC3O]

Figure 7.7 — The shield is soldered to the outside of the rear of the connector body and will be wrapped with tape while still warm to provide a seal. [ROBERT BASTONE, WC3O]

UHF Plugs with Smaller Cable

83-1SP (PL-259) Plug with Adapters (UG-176/U or UG-175/U)

1. Cut end of cable even. Remove vinyl jacket 3/4" - don't nick braid. Slide coupling ring and adapter on cable.

2. Fan braid slightly and fold back over cable.

3. Position adapter to dimension shown. Press braid down over body of adapter and trim to 3/8". Bare 5/8" of conductor. Tin exposed center conductor.

4. Screw the plug assembly on adapter. Solder braid to shell through solder holes. Solder conductor to contact sleeve.

5. Screw coupling ring on plug assembly.

ARRL0908

The PL-259 can easily be adaped to smaller cable types such as RG-58, 59 or RG-8X through the use of an adapter (some call it a reducer) that fits into the threaded portion of the plug. In many ways, this is easier to accomplish than the connection of the larger cables, even though the steps are similar. The cable is somewhat easier to handle and easier to strip since the dielectric is thinner and doesn't need to be cut through in one step. The shield also has less thermal mass to deal with while soldering. The process is illustrated in **Figure 7.8**.

Just as with the coupling ring and the larger cables, it is very important to slide both the coupling ring and the adapter onto the cable before progressing very far. It is also very important to support the connector and cable during soldering so it won't bend and allow the center conductor to migrate through the dielectric.

In Step 3 of Figure 7.8, it is important to trim the shield so that it doesn't extend over the threaded portion of the adapter; otherwise strands will jam the threads. Following Step 4, after the connector cools, make sure that the shield is soldered to both the adapter and the connector body. If it isn't properly soldered, you will be able to turn the adapter in the body. If that happens, just reheat the rear of the connector body and apply more solder into the holes when it's ready to flow. As with any connector installation, check for shorts with an ohmmeter.

Figure 7.8 — The procedure for assembly of UHF plugs with smaller coaxial cable. The UG-175/U adapter is for RG-58 size cable with 0.195-inch outer diameter. The UG-176/U adapter is for RG-59 size cable with 0.242-inch outer diameter, including RG-8X and RG-62. [AMPHENOL ELECTRONICS, RF DIVISION]

UHF Plugs with Crimped Connections

PL-259 compatible connectors are available with crimped rather than soldered connections for both large and small cable sizes. Each requires different size ferrules and different crimping tools. Some types have a center conductor that is soldered while the shield is crimped, while others have crimp arrangements for both conductors. If you use a type with a crimped center pin, make sure the center conductor extends all the way to the bottom of the pin, or the connection won't be solid. (See **Figure 7.9** for assembly details.)

UHF Sockets

The standard SO-239 panel socket mated with the PL-259 is a flange mount arrangement with a center pin and only four mounting holes for shield connection. The usual practice is to mount the connector using four machine screws, lock washers, and nuts. This provides a connection to a metal panel. If it is desired to continue with coax inside the panel, the usual practice is to connect the shield, via as short a connection as possible, to a solder lug under one (or more) of the mounting nuts.

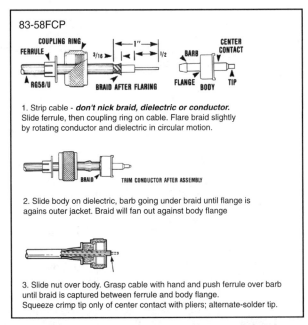

Figure 7.9 — The procedure for assembly of one type of crimped UHF plug with smaller coaxial cable. The 83-58 family of connectors are available in multiple versions for different cable types with different ferrule diameters requiring different crimp dies. Some have a soldered center conductor, some (as shown) have an inner conductor that can be crimped or soldered. [AMPHENOL ELECTRONICS, RF DIVISION]

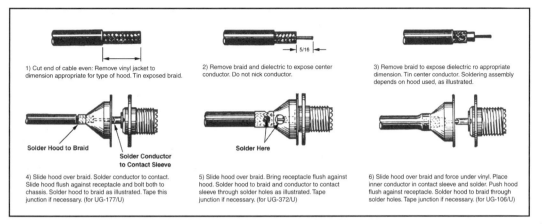

Figure 7.10 — The official procedure and dimensions for assembly of a shielding hood for SO-239 flange mount jacks. The different part numbers are for different cable types.

This arrangement is obviously not satisfactory if high frequency shielding is required. Using a shield hood can result in a properly shielded connection. (Details are shown in **Figure 7.10**.)

In addition to flange mounting, one-hole mount sockets are available. These are constructed much like the threaded PL-258 barrel connectors and are secured with nuts that fit the backshell threads. A center pin solder connection is provided inside the panel and if a shield connection is required, it is often made using a large solder lug that fits around the connector body. To extend a shielded connection with a one-hole mount, a PL-258 type double female connector with matching nuts can be used. The coax inside the panel is then terminated with a PL-259 plug.

INSTALLING TYPE N COAXIAL CONNECTORS

The Type N coaxial connector series is superficially similar in size and operation to the UHF series, but it corrects the major deficiencies of the earlier UHF type:

♦ Unlike the UHF, the Type N connectors maintain a constant characteristic impedance through the connector.

♦ The construction is such that the shield connection is not dependent on a tight backshell. Of course, if the backshell is removed far enough, the connectors can be pulled apart and break contact.

♦ The gaskets of the Type N prevent water penetration, a major factor in the deterioration of coax cables.

A Type N plug is actually easier to assemble than the UHF counterpart.

So why don't more amateurs switch over to Type N? Simple: They are more expensive (typically around $6, compared to $2 for a high-quality plug) and many installers don't know how to deal with them. The more expensive part is relative — what's it worth to save a trip up the tower in the winter?

Select the Correct Connector

Type N connectors come in 50 and 75 Ω versions. Be sure to select the right series, and get the right connector for your cable type if you are concerned about maintaining impedance. The insidious problem with having 50 and 75 Ω versions is that they look identical, and can be made to mate with each other, to a degree. As with the RG-8 (50 Ω) and RG-11 (75 Ω) type cables themselves, the connectors have the same outside and shield contact diameters. In order to maintain the constant impedance, the 75 Ω N connectors have a smaller diameter center conductor pin.

If a 50 Ω plug is inserted into a 75 Ω socket, the socket's inner conductor will be spread open too far, making it unreliable for use any later use with its proper 75 Ω a plug. While inserting a 75 Ω plug into a 50 Ω socket will not cause damage, it will result in an uncertain center conductor connection. If the pin is in perfect alignment, there will actually be no contact at all.

Installing a Type N Plug or Socket

Unlike the standard UHF connector arrangements that just have cable plugs and panel jacks, the Type N is available with similarly installed cable plugs and cable jacks. This makes coaxial "extension cords" feasible without the need for a double female adapter.

The installation instructions for a Type N plug or cable socket are shown in **Figure 7.11**. To add clarification here, we expand upon these steps using an Amphenol connector, part number 82-202-RFX, designed for a number of 0.405-inch diameter 50 Ω coax cables, including RG-8 and RG-213. Note that the suffix RFX indicates it is a commercial version rather the 82-202, which is the full military type at about three times the price. (Visit their website at **www.amphenolrf.com** to find the right kind for your cable.) A step-by-step installation process follows.

♦ *Lay it out* — Slide the nut, washer, and gasket over the cable jacket — in the correct order.

♦ *Remove the outer jacket* — Measure 9/16 inch of the outer jacket and carefully cut and remove it with a razor knife, without nicking the braid.

♦ *Comb out the braid* — Using an awl or a metal comb, straighten out the shield wires and slide on the clamp until it stops on the jacket, as shown in **Figure 7.12**.

♦ *Strip the dielectric* — Remove enough of the dielectric to leave

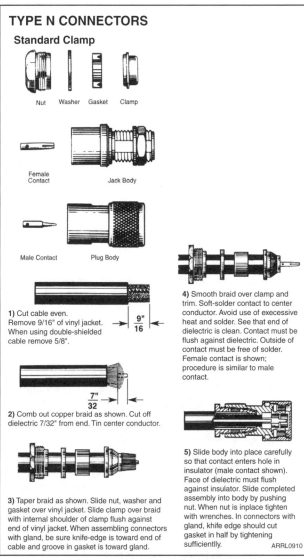

Figure 7.11 — The official procedure and dimensions for assembly of a standard Type N plug or cable jack.

Figure 7.12 — The nut, washer, and gasket are in place on the cable jacket, in the correct order. The right amount of outer jacket is removed, and the shield wires are combed out straight.

$7/32$ of an inch between the end of the dielectric and the end of the inner conductor.

♦ *Trim the shield wires* — Use sharp diagonal cutters to trim the shield wires to make them just the right length to cover the round portion of the clamp (**Figure 7.13**).

♦ *Tin the center conductor and solder the pin* — Lightly tin the center conductor. Check to make sure it fits in the pin, and lightly dress it with a fine file, removing enough solder so that it fits fully into the pin. The bottom of the pin should rest against the dielectric and, with the pieces pushed together (as shown in **Figure 7.14**), the pin should be flush with the end of the housing. Now is the time to adjust as needed, then solder the pin and remove any solder on the outside of the pin with the file or emery cloth.

♦ *Final assembly* — Insert the pin into the connector body dielectric by pushing on the cable. Turn the nut until the threads start. Hold the connector

Installing Coaxial Connectors on Cable

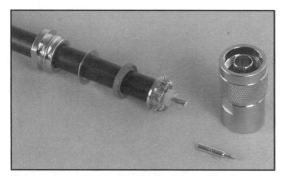

Figure 7.13 — The shield wires are trimmed to fit over the round edge of the clamp and enough of the dielectric is removed to allow the pin to fit flush.

Figure 7.14 — Lightly tin the center conductor, make a final check of dimensions, and then solder on the pin.

Figure 7.15 — The fully assembled connector.

body with a wrench and turn the nut with another wrench until tight. The pin should be about flush with the end of the coupling, the connector should not easily turn on the cable, and you should not be able to pull it off the cable (**Figure 7.15**).

Type N Jacks

As previously noted, the Type N offers cable jacks that assemble in the same manner as cable plugs. They also offer panel jacks in a few configurations. There are types that are similar to the open frame SO-239 panel jack, but there are also panel jacks that maintain the characteristic impedance and shielding integrity, more appropriate for critical RF applications. These assemble in the same manner as the other plugs and jacks, and are best installed in the panel after assembly.

Some Observations

Note that with the Type N, we have avoided the step of soldering to the connector body. This avoids the risk of melting the cable dielectric, a frequent problem with UHF connectors. Also, it is possible to modify the procedure slightly and solder on the pin, push the cable with pin attached through a small bulkhead hole, and then assemble the connector on the other side. With a UHF PL-259 connector, you need to do the soldering on the far side of the wall.

Unless there is a lot of working room, it is not recommended to replace UHF connectors on equipment with Type N. Adding a UHF to Type N adapter to a UHF connector, however, especially when the UHF is still new, can avoid some of the typical problems of loose coax connections later on.

For cables of the RG-58, 59, and 8X variety, the BNC connector is a good choice. It offers many of the benefits of the Type N in a smaller, bayonet-attached package. Type N connectors are also available for virtually all types of coax the amateur is likely to encounter, although they will be harder to find and may not be available in the less-expensive commercial series. A BNC connector with a BNC to N between series adapter may be a good solution.

Type N Plugs for the Dedicated UHF Plug User

It is no wonder that hams are wedded to the ubiquitous PL-259 UHF series plug, since most HF and a large portion of VHF equipment provides matching UHF sockets for connection to antennas. However, there is a line of Type N plugs that assemble in almost the same way as a PL-259 plug. In fact, as seen in **Figure 7.16**, the shield is connected in exactly the same way, whether the cable is RG-8 or RG-8X. The same UG-176 adapter for RG-59 or similar sized coax that fits a PL-259 also fits in this Type N connector, as does the UG-175 for RG-58.

The same techniques that are used to prepare either size cable for a PL-259 are also used here. The same tools, such as the Ripley UT-8000 jacket and dielectric stripper, can be used to prepare the larger cables, or the traditional knife and pliers approach will work as well for this connector.

The connector comes in two pieces: an inner body where all the soldering takes place that includes a captive pin, and a housing that includes the backshell and the shield connection arrangement. As with a PL-259, this Type N can be disassembled, cleaned out, and reused. The usual Type N includes a gasket that is destroyed upon first assembly, and those can't be found either.

With the backshell removed, the soldered connector interior has a diameter of 0.6 inches, significantly less than assembled Type N at 0.83 inch,

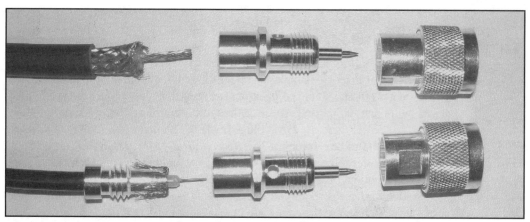

Figure 7.16 — RG-213 and RG-8X cable about to be assembled into UHF-like Type N plugs.

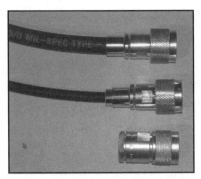

Figure 7.17 — Fully assembled Type N coax connectors on RG-213 and RG-8X cable. A standard clamp-type Type N plug is shown for comparison.

or even PL-259 at 0.74 inch. This can make it easier to route cables through smaller holes in partitions. The standard Type N can be soldered to its pin and routed through holes as small as the cable itself, but the final assembly has to be completed on the far side of the partition. Of course the current connector also needs final assembly, but it's just the threaded backshell piece. This can easily be tightened up using two $\frac{9}{16}$-inch open-end wrenches, or one wrench and a vise.

The only difference between assembling this connector and a PL-259 is that the center conductor wire has a maximum length before it bottoms out inside the center pin. It must be trimmed until it can be screwed in far enough that the shield appears beyond the solder holes, as the connector is screwed onto the coax or the UG-175/6 adapter. **Figure 7.17** shows the assembled connectors in comparison to the size of a standard solder and clamp Type N plug.

These connectors cost between $5 and $7, about twice what a quality PL-259 costs, but about the same as the standard solder and clamp Type N plugs. They are somewhat longer than the usual Type N, which may be an issue in some installations. The standard N plug is about 1.5 inches long, while the item under discussion is 1.785 inches, 1.96 with a UG-176 adapter.

INSTALLING BNC AND OTHER COAXIAL CONNECTORS

Each coaxial connector type has its own installation requirements, although the BNC is quite similar to the Type N connector described in the last section.

Installing BNC Connectors

Installing a BNC connector is very much like installing a Type N, except it is mostly smaller. While the assembly is much more compact, the center pin and shield connections are the same size. In fact, a Type N plug will mate with a BNC socket although the coupling ring will not function.

While there are BNC plugs made to terminate the larger cable sizes, they tend to be a force fit. The BNC is ideally suited to RG-58 and RG-59 (or smaller) size cables and makes a very handy patch and interequipment arrangement because of its quick-change bayonet-secured backshell. If you stick to these sizes of cable, the bayonet is even easier to assemble than a Type N because the cables are much easier to work with.

As with others, the BNC is available in both crimp and clamp styles.

BNC CONNECTORS

Standard Clamp

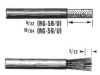

1. Cut cable even. Strip jacket. Fray braid and strip dielectric. **Don't nick braid or center conductor.** Tin center conductor.

2. Taper braid. Slide nut, washer, gasket and clamp over braid. Clamp inner shoulder should fit squarely against end of jacket.

3. With clamp in place, comb out braid, fold back smooth as shown. Trim center conductor.

4. Solder contact on conductor through solder hole. Contact should butt against dielectric. Remove excess solder from outside of contact. Avoid excess heat to prevent swollen dielectric which would interfere with connector body.

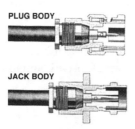

5. Push assembly into body. Screw nut into body with wrench until tight. **Don't rotate body on cable to tighten.**

ARRL0911

Figure 7.18 — The official procedure and dimensions for assembly of a standard BNC plug or cable jack.

The solder (pin) and clamp (shield) are so easy to do with smaller cable, and are recommended if you can read a ruler and have power for a soldering iron. A notable exception would be cables with foil (or mostly foil) shields. The foil is somewhat less suitable for the shield clamp and a good match with a crimp arrangement.

Figure 7.18 provides the step-by-step instructions for BNC assembly. As with the Type N, both cable and panel jacks are available that assemble in the same way. There are open-frame jacks available in both flange type and one-hole mount styles.

Beware of electronic retailer "BNC" plugs that have a BNC-size backshell but use the center conductor of the cable as the pin. While they advertise them as "no solder" types, the wire is not the correct size to mate with a BNC jack.

Type F Connectors

Probably the most common coaxial connector ever made is the Type F, described in Chapter 6. While rarely encountered in Amateur Radio equipment, it appears in virtually all television and consumer video equipment made since the 1970s. It may be encountered in amateur video, or in making use of cables originally intended for other services. Because of its high volume, and the nature of the consumer marketplace, it was designed to be inexpensive and easy to install. In spite of this, cable systems using the connector must meet stringent leakage standards because cable systems use most of the spectrum allocated to over-the-air services and they must not interfere with them.

The Type F plug is designed for use with RG-6 size cable, somewhat larger than RG-59 and generally used from the pole to the subscriber, and smaller sizes used within a premise or between equipment. A screw-on backshell on the F plug maintains shield contact with the barrel of the socket, but as with the PL-259, it is only effective if tight. There is also a "push-on" type plug available that uses spring fingers to grasp the jack barrel. The solid (only) inner conductor of the coax pushes into a spring contact on the jack, generally with significant through clearance so the length is not critical, unless it doesn't make the spring contact.

Cable TV-type flexible coax provides a large fractional shield coverage through the use of a wrapped aluminum foil shield, sometimes in combination with a loose wire shield. This arrangement pretty much rules out soldering of the shield connection and provides either a clamp or, more often, a crimp-on connection. Again, a number of different proprietary types of plugs with proprietary tools are available for such applications as waterproof outdoor use, although the very basic types are not very fussy about the crimp tool used. Still, a good fit makes for a better and neater connection.

RCA (Phono) Connectors

The RCA connector (see Chapter 6) was originally employed to connect a low-level phonograph cartridge signal, via a coax-like shielded wire, to an amplifier. As such, it has an inherently coaxial structure with the plug having a hollow pin and the shield connected via four springy petals, making a non-threaded almost, but just barely, backshell.

The connector type is most frequently encountered in Amateur Radio applications as a between-equipment connection arrangement for auxiliary functions, such as carrying reference and oscillator signals as well as switching, control and sometimes low voltage dc lines. In the 1950s and 1960s, it was sometimes also found used in place of a UHF connector as an antenna connection on receivers and transmitters at or below 100 W PEP.

While crimp-type connectors may be available, loose connectors tend to be the solder type. The basic connector consists of a center conductor soldered within the hollow center pin and the shield soldered over the outside of the shield fingers in K3LR fashion. Such connectors provided as part of preassembled audio cables with molded housings may be of a different arrangement.

In addition to the basic type, there are some with a metal screw-on back cover and interior terminal for the shield. These are neater and easier to grasp, but don't offer any electrical advantages over the basic configuration.

Chapter 8

Determining Which Line is Best Suited for a Particular Application

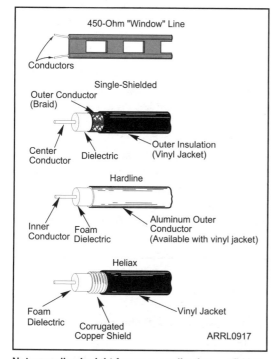

Not every line is right for every application — picking the right one may offer improved performance and longer cable life.

The choice of the best transmission line for an application may make a big difference in how well the system works. Some application problems appear immediately, sometimes with dramatic results, while other, more insidious ones manifest themselves gradually over time and can be easily overlooked.

GETTING THE SIGNAL TO THE FAR END OF THE LINE

What we usually want most from our transmission line is to get the signal to come out the far end looking as much as possible like the signal with which we started. In the realm of radio signals, this statement is mostly about transmission line loss. In other applications, such as pulsed data systems, we may have other concerns, including preserving pulse shape without distortion, but here we will focus on transmission of RF energy. As we have discussed, the selection of line type plays a crucial role in how successful we will be.

We spend a lot of money to have a transmitter to generate a radio signal, and we want that signal to arrive at our antenna and leave as radiated RF energy — not be radiated as heat from our transmission line. There are many factors that contribute to line loss, most of which we have already covered in detail. Here we will discuss which is most important for a particular application.

Characteristic Impedance

The characteristic impedance (Z_0) of a transmission line is one of the most fundamental attributes of a transmission line, and often considered first in selecting a line. While the Z_0 does relate directly to loss (higher Z_0 lines generally have somewhat less loss, all other things being equal), the matched loss due to the Z_0 is usually less important than the extent to which the Z_0 matches the load.

The Matched Case — In the most common configuration, the load is of approximately the same impedance that the equipment is designed to drive. That impedance is most often 50 Ω for radio equipment and 75 Ω for video equipment. Not coincidentally, the most common coaxial transmission lines are available with those Z_0 and the obvious choice for Z_0 is the one that matches the actual Z impedance of the equipment at both ends. Any other Z_0 will result in an SWR greater than 1:1, which will increase cable loss and not properly load the equipment.

Once we know the Z_0, we have a number of other considerations — how much loss will our system design tolerate, and what power rating does the cable need to meet. **Table 8.1** shows representative data on these parameters for some of the most popular 50 Ω cables. Note the dependence on operating frequency for both of these parameters.

If losses are of a particular concern, it is possible to change cable types to meet particular requirements. For example, if the load is a rotary antenna, it is usually necessary to feed it with a loop of flexible cable to avoid fatigue as the rotator turns back and forth. Since the lowest loss cable types do not tend to be very flexible, it is possible to change types (within the same Z_0) if the run is long enough to make a difference. At most frequencies, there is

Table 8.1
Attenuation and Power Handling Capability for Various 50 Ω Coax Cables vs Frequency

Cable	Frequency (MHz)	Attenuation (dB/100')	Power Rating (W)	Minimum Bend Radius (")
RG-58*	10	1.2	1937	2
	100	4.3	542	
LMR-195**	10	1.1	1430	
	100	3.6	450	
RG-8*	10	0.6		4.5
	100	1.9	800	
LMR-400**	10	0.39	5800	
	100	1.2	1810	

*Belden data from **www.belden.com**
Times Wire and Cable data from **www.timesmicrowave.com.

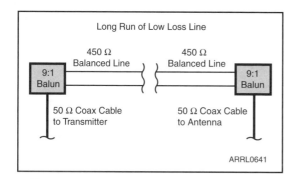

Figure 8.1 – The transition from coax to low loss, air dielectric, balanced transmission line can reduce loss if the loss reduction in the line is greater than the loss in the baluns. Here 9:1 (450: 50 Ω) baluns are used to transform the 50 Ω end impedances to the 450 Ω impedance of the window line.

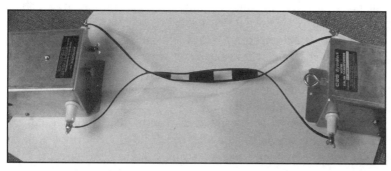

Figure 8.2 – Physical model of the 9:1 baluns from Figure 8.1 and a short window line section set up for laboratory loss testing.

negligible additional loss due to the connectors required; however, they do become additional points of potential failure as well as points of possible water ingress, if precautions aren't taken.

It is also possible to change to a very low loss medium if the length is sufficient. For example, the loss of 450 Ω window line, or 600 Ω open-wire line, is significantly less than that of coax, particularly as the frequency goes up. A balanced to unbalanced impedance transformer (balun) can be used to transform the impedance of the low loss medium to the desired 50 Ω at each end of the run as shown in **Figure 8.1** and **Figure 8.2**. The trade-off here is that the transformers will have some loss.

ARRL Lab measurements during the experiment pictured conclude that the pair of HF baluns had a total loss of less than 1.0 dB. Thus the run needs to be long enough so that the difference in loss between the cable types is greater than 1.0 dB in order for this to make sense. It is also necessary that the balanced line be installed so that it is away from the ground and not subject to other interfering objects.

Mismatched Case — If the transmission line isn't matched to the load or the equipment, a number of additional considerations need to be addressed. The three main areas are:

♦ *Additional Cable Loss.* As noted in Chapter 3, a mismatched transmission line will have more loss than a matched one. This is not generally a problem with a reasonable mismatch, perhaps 2:1 or 3:1, but it all depends on how high the matched loss would have been for the cable type, length and frequency.

♦ *Equipment Limitations.* Many radio transceivers will engage foldback circuitry if the load is different than that specified. This circuitry is designed to protect the output stages of a transmitter from the higher voltages or currents that could be present in a mismatched environment. While some radios are happy to drive loads with 2:1 or higher SWR, some start reducing power at 1.5:1. In terms of getting a signal to the end of the cable, this can be even more significant than cable loss.

Some types of equipment that follow a linear source impedance model will also deliver less power to the cable with a mismatch than if it is matched. The answer to these issues is to understand the requirements of your equipment and translate the equipment specifications into requirements for your transmission line system.

♦ *Stress on Components.* A standing wave ratio greater than 1:1 implies a changing current and voltage along the line. This means that the voltage and the current will be higher and lower along the length of the line than they would be if the line were matched. Fortunately, the locations of high current will be the locations of low voltage, and vice versa.

If the complex impedance of the load and the line propagation velocity and length are known, it is possible to determine the actual voltage and current at the end of the line. However, these values will hold for a single frequency, so in a multiband system, each may appear anywhere. The maximum voltage and current will each be equal to the values in a matched system times the square root of the SWR. If the termination equipment is specified to handle the resulting value, there should be no problem. As an example, with a matched 50 Ω, 1500 W system, the line voltage will be about 274 V, the current about 5.5 A. With a 4:1 SWR, the maximum voltage and maximum current will be twice those values at some points along the line.

OUTDOOR ISSUES

Generally, all modern transmission lines can be successfully operated indoors or outdoors. There have been types of TV twinlead, available years ago, specifically intended for indoor use because they were less visible. After a few seasons in sunlight, they would crack and eventually disintegrate. While modern cables last much longer outdoors, they do suffer degradation to different extents due to sunlight (primarily UV radiation) and moisture. The main degradation mechanism is the breakdown of the outer jacket, the rate of breakdown being dependent on the characteristics of the jacket material.

Jacket Contamination

As part of the jacket manufacturing process, plasticizers and other compounds may be left within the jacket material. These materials can leach and migrate from and within the cable, changing the characteristics and causing degradation of both the shield and the inner dielectric. This process is accelerated by the temperature rise resulting from sunlight exposure. Cables identified as *non-contaminating*, such as those with non-contaminating vinyl (NCV) jackets, are resistant to these chemical effects. High-quality PVC and other materials can provide good protection to both UV and high temperature degradation mechanisms.

Polyethylene is another jacket choice that does not have plasticizers and is inherently resistant to UV, as well as offering additional abrasion resistance. Good-quality PVC jacketed cable can be expected to last 9 to 14 years before it suffers significant degradation, while polyethylene jacketed cable can last about twice as long. Low-quality cable can be expected to degrade significantly faster.

Water Penetration

Coaxial cable jackets are made of different materials suitable for different environments. A major problem with many types of coax, as well as some connectors on the ends, is water penetration. Water within the coax can cause the dielectric to become lossy and result in corrosion of the shield and even the inner conductor.

Water enters coaxial cable in at least two ways. Gaseous water vapor can migrate through the outer jacket. Liquid water can also enter through jacket pinholes left from the manufacturing process, or from abrasion resulting from rough handling.

In most cases, the entry of water into the cable from the connector ends is a more significant issue. While BNC and Type N connectors are designed to be waterproof, PL-259s are not. In order to avoid problems, a PL-259 must be sealed by means external to the connector. In addition, it is particularly critical that a drip loop be included at each end of every cable run that goes downward. A drip loop (see **Figure 8.3**) is any local low point before the connector so that water running down the outside of the cable can drip or run off before it gets to the connector. Without the loop, water will pour onto the connector, sometimes coming right into the equipment if connections aren't tight.

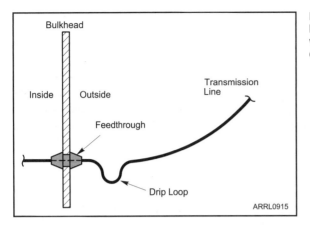

Figure 8.3 – A drip loop is an intended local low point in the cable before the connector so water running down the outside of the cable can run off before it gets to the connector.

There are a number of commercial products available for waterproofing connectors, although some make later connector removal very difficult, so make sure the connection is right before you seal it. High-quality cable will have fewer pinholes and be less subject to vapor migration, but improper connector installation can negate the benefits.

Direct Burial Coax

Most types of coax are designed for above-ground use, but some are rated for direct burial. This is mostly a statement about abrasion resistance. The usual ground near the surface moves with seasonal changes as well as with differences of surface load. Any shifting of the ground will tend to result in jacket abrasion from rocks or other debris in the soil.

Note that direct burial does not mean "submersible." Only special submarine cables, out of financial reach for most amateurs, are designed to spend time in the water. Direct burial cable doesn't promise lack of pinholes. This can result in problems for cables in any kind of ground without good drainage, such as clay. If the older houses in the area are made of brick, check your soil carefully before you bury any coax cable.

Some installers use conduit to avoid abrasion problems with below-ground cable. While this can reduce abrasion, it can result in other problems, such as water accumulation. Water can enter conduit through the ends or, more likely, through condensation. Without proper drainage, the coax ends up in the water. To avoid this problem, implement one of the following:

◆ Use conduit or drainage pipe with drainage holes on the bottom side. Prepare the trench with a layer of gravel above sand or earth with good drainage. Lay the conduit in the trench, hole-side down, and use elbows on the entrance and exit points to avoid direct water entry (see **Figure 8.4**).

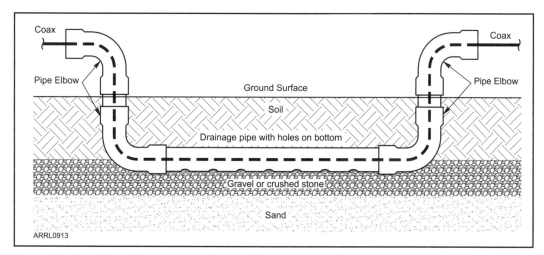

Figure 8.4 – A problem with burying coax in conduit is the accumulation of water in the conduit. By using drainage pipe with holes on the bottom side over good drainage material, such as the gravel and sand shown, any water can be moved away from the cable.

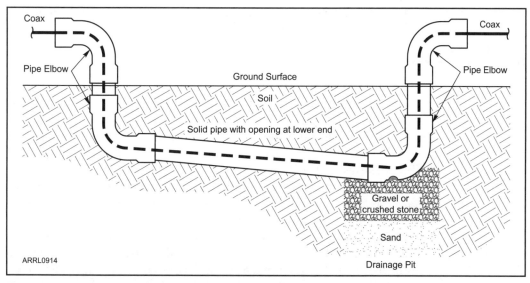

Figure 8.5 – Another solution to the problem of accumulation of water in the conduit is to use a pitched conduit with one or more low points able to drain into drainage pits.

◆ Lay the conduit with a pitch so that all water will run to one end, or to a midpoint. Have a drain at that point into a pit with gravel and sand in sufficient quantity to absorb the expected water. Follow up with a look at the drainage pit after some wet weather (see **Figure 8.5**).

Determining Which Line is Best Suited for a Particular Application   8-7

Window Line

Because the fields associated with signals traveling on window line are largely outside the transmission line, the line is affected by anything within a few widths of the line. Such interaction can have two distinct results. Balanced line near metal takes on the characteristics of a double stripline with a resulting change in Z_0. Depending on application, this may be significant, particularly if a system intended to be matched. Of generally more serious consequences, window line will experience considerable loss if placed close to a lossy medium, such as on the ground. Our experiments indicated that the loss in 100 feet of window line on dry ground was 10 to 20 dB over the HF region. This is something to watch out for, especially during temporary operations in which normal installation standards may be neglected.[1]

UNFRIENDLY RF AND MATERIAL ENVIRONMENTS

To a greater or lesser extent, all transmission lines are subject to effects of their external environment. Some of these are obvious. For example, a transmission line that melts in a high-temperature environment will not work at specification very long, but there are more subtle interactions with the world outside the line.

RF Environment

The ideal transmission line is one that sends signals from one point to another without any interference from or to what's in between. No transmission line is quite perfect in this regard, but some are more susceptible than others. This effect is bilateral as well — the signals we send are supposed to stay inside the line, but they don't always do so. There are three primary mechanisms at play in this realm.

Proper Line Termination. We have already discussed matching impedance as a termination issue, but that isn't a factor here. To consider signals getting outside the transmission line, we consider the balance of the terminating load. An unbalanced line, such as coax, needs to be terminated by an unbalanced load. If instead it is connected to a balanced load, perhaps a center-fed dipole, the current from the inside of the coax will split between the half-antenna connected there and the outside of the coax, acting like another wire in parallel.

The amount of current going on the outside of the coax, which — because of skin effect — acts like a separate conductor from the inside surface of the shield, depends on the relative impedances of the two connection paths. This will depend on the frequency, the length of the coax, and the impedance of the ground termination of the shield at the far end.

Any current on the outside of the coax will result in radiation, just as from an antenna. Depending on the system design, the environment, and the sensitivity of other equipment close by, this may or may not be a problem — and in some cases, it can be an advantage. If the coax is connected to a transmitting antenna, and it runs past alarm system wiring in the building, it can be a serious problem. If you spent lots of time tuning a Yagi antenna for best front-to-back ratio and the coax picks up signals from all directions, you will also not find it a plus. This can be a problem in either direction, and is often the source of radio frequency interference (RFI) from household appliances as the coax works its way to the station.

Balanced transmission line, such as the popular window line, suffers from the same issues. It is designed to feed a balanced load, such as that center-fed dipole. If the load is not balanced, perhaps due to sides of the dipole being at different heights or near different objects, the currents on the two conductors will not be equal and opposite. Any resulting difference current due to imbalance will act like a separate antenna current on the line, and radiate just as will the coax described above, resulting in all the same issues. In addition, the unshielded balanced line, while its balanced currents cancel at distance, even if perfectly balanced can couple to conductors or objects near its path if the object is closer to one conductor than the other.

Coax Shield Coverage. The ability of coax cable to keep signals within the medium depends in no small measure on the integrity of the medium — in this case the shield. For coax to operate perfectly in this regard, it requires a perfect shield. This would be an infinitely thick, zero resistivity-conducting medium surrounding the dielectric, which unfortunately isn't available for purchase. Some coax, such as cable TV distribution cable, which runs between poles down a street, has a shield of solid aluminum tubing that comes pretty close, but it takes a tubing bender to make a nice corner — not what the typical amateur wants for many applications, but great if you want a long, low-loss, 75 Ω coax run out to your antenna field. Cable companies make use of, and must avoid interference to, frequencies assigned for over-the-air use by other services. Government agencies get particularly upset if there are signals radiating on aircraft navigation and communication frequencies from poorly shielded coax, so cable TV companies have been sensitized to this issue.

Most coax used by amateurs has a braided copper shield with an effectiveness related to the percentage of coverage — often a parameter listed by manufacturers. Note that even 100% coverage by braid will not quite equal the shielding effectiveness of tubing. Still, all things being equal, more coverage means more effective shielding. The 75 Ω RG-6 cable, often used as "drop cable" between the street and an end user, typically has a dual shield

— one layer aluminum foil, the other tinned copper braid. While the braid is usually quite sparse, the combination meets the strict cable TV emission requirements. Unfortunately, it doesn't work well with soldered connectors, but is appropriate for crimp-on types.

Connector Continuity. Particularly with UHF and F-type connectors, in which the connection of the shield is dependent on connector backshell tightness, improper tightening of connectors can nullify part of the effectiveness of shield coverage. Any resistance in the shield connection acts like a coupling mechanism between the inside and outside of the shield. This can get worse with time, due to vibration or oxidation of the surfaces.

Material Environment

Coax cable pretty much has the advantage here. The fields between the conductors of coax, except as noted above, stay largely within the cable itself. Thus coax is pretty impervious to normal outside influences — other than water or abrasion as outlined previously. Still, each cable has a temperature rating that should be observed, as well as some sensitivity to some types of chemical vapor environments that can attack the outer jacket or permeate and degrade the shield or inner materials.

Window line, with most of its fields outside the physical location of the wires, is very susceptible to signal degradation of various types due to proximity effects. This should be carefully considered in selecting appropriate line for a particular application. As noted previously, the attenuation of window line closer than a few inches from dry ground is very high — perhaps 10–20 dB in about 100 feet, based on our measurements.

If window line is routed close to metal surfaces, it will have a major change in characteristic impedance, as well as the possibility of coupling signals to undesired objects. Similarly, window line can be a problem in getting through walls or other partitions in which the lossiness of the materials in the partition may be unknown. One solution may be to use short sections of dual coax (see Figure 5.9 in Chapter 5); however, they should be quite short if the coax loss is to be negligible.

Notes
[1]B. Allison, WB1GCM, J. Hallas, W1ZR, "Getting on the Air — A Closer Look at Window Transmission Line," *QST*, Nov 2009, pp 66-68.

Chapter 9

Application and Installation Notes

Coaxial cables going up a tower at W1AW, the ARRL Headquarters.

Once you have decided on a transmission line with the features you want, it is important to install it properly. Poor installation practices can result in more rapid degradation of transmission lines due to additional stress or inappropriate environmental factors. Implementing some of the following elements can help extend the life of your transmission line.

SUPPORTING LINE RUNS

Transmission lines are not generally intended to be structural members, yet they are often treated as such. A transmission line, particularly a coaxial cable, is not light and, in typical lengths, can be fairly heavy. **Table 9.1** provides some of the physical characteristics of typical coaxial cables. A quick glance indicates that each type should be able to withstand the tension of at least a 1000-foot vertical run; however, that's not quite the whole story.

Vertical Transmission Line Runs

While the cable itself may be able to withstand the tension of its own weight, whatever is used to secure the upper end will need to exert some kind of force on something to help it defy the forces of gravity. While a connector should also be able to support the weight, (assuming proper installation and no vibration that will cause it to break free in time), hanging a long line on a connector is not generally a great idea. The only thing worse is to have everything hanging directly on the soldered connections between the antenna and the cable.

Table 9.1
Physical Properties of Selected Transmission Line Types*

Cable Type (Belden PN)	Weight/1000 feet (pounds)	Max Tension (pounds)	Minimum Bend Radius (inches)
RG-213U (8267)	104	184	4.0
RG-8U (8237)	104	190	4.5
RG-8U Foam (9913)	97	300	6.0
RG-8U Foam (8214)	106	230	4.0
RG-8X Foam (9258)	35	75	2.4
RG-58 (7807)	22	25.4	1.9

*www.belden.com

The usual clamping arrangement will exert perpendicular forces that will distort the shape of the cable and result in a combination of concentration of tension forces and reduced ability to withstand the tension.

One good way to provide support to a coaxial transmission line is shown in **Figures 9.1 through 9.5**. This was described by Lyle Nelson, ABØDZ, in a *QST* "Hints and Kinks" column.[1] This method distributes the tension along a length of the cable and doesn't require any tight clamping to hold it securely in place. To further reduce tension on the line, this may be repeated as often as desired so each section of the cable is just holding up itself.

Horizontal Transmission Line Runs

A horizontal run adds the force required to avoid sag in the run. In fact, strictly speaking, it is only possible to have a horizontal run if there is infinite tension on each end to result in zero sag. All practical runs over horizontal spans are actually in the shape of a catenary, the shape that a cable takes on if its weight is supported at each end (see **Figure 9.6**).

To determine the amount of tension needed for a particular amount of sag, use the nomograph shown in **Figure 9.7**. The operation is performed in two steps:

Step 1. Lay a ruler between the two outside scales, with one end on the desired span length, and the other on the cable weight (in the same units as in Table 9.1). Make a dot on the "Work Axis."

Step 2. Rotate the ruler, keeping an edge on the dot from Step 1. At each position there will be a value for the tension and the corresponding half-span sag. Use a tension that is comfortably below the maximum tension shown in Table 9.1.

Figure 9.1 – An attachment mechanism for securing coaxial cable that distributes the tension of a length of the line to avoid excessive stretch. This can be used for either horizontal or vertical cable runs. The contributor recommends a 40-inch length of ⅛-inch braided nylon rope for RG-8X size cable, 60 inches for RG-8 size, with an overhand knot at each end and the ends melted to avoid fraying. [LYLE NELSON, AB0DZ]

Figure 9.4 – Wind turns until you have about 3½ inches of the loop covered. [LYLE NELSON, AB0DZ]

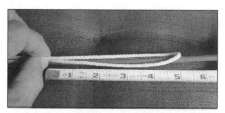

Figure 9.2 – Start by making about a 5-inch loop along the cable. [LYLE NELSON, AB0DZ]

Figure 9.3 – Wind the cable in tight wraps starting at the open end of the loop. [LYLE NELSON, AB0DZ]

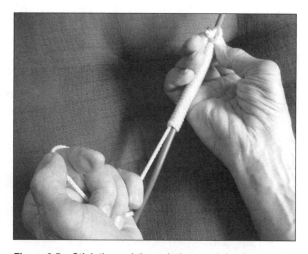

Figure 9.5 – Stick the end through the remaining loop and pull the rope from the other end until you have it tightly secured. The "tail" left out is then used to tie the cable to a screw eye or other support. [LYLE NELSON, AB0DZ]

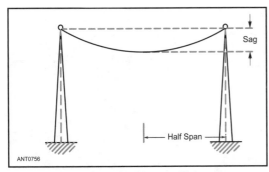

Figure 9.6 – A horizontal cable run with less than infinite tension on the ends will assume the shape of a catenary, as shown.

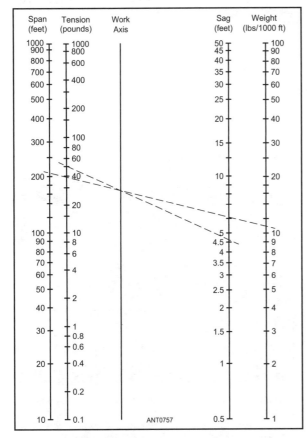

Figure 9.7 – Nomograph for determining sag in a cable run based on cable weight and span distance.
[JOHN ELENGO, JR, K1AFR]

The dashed lines on Figure 9.7 are for an example with a cable weight of 11 pounds per 1000 feet and a span of 210 feet. The second line indicates that for a tension of 50 pounds, a sag of 4.7 feet will result.

Note that all of this assumes the end supports are fixed. If the end supports are trees, or other objects that will shift in the wind, the system needs to be laid out as if the trees were both at the position of furthest separation. The result will be additional sag in the rest position.

If the resulting sag is not satisfactory, or if you wish to reduce physical stress on the cable, a *messenger cable* can be used to support the cable run. This is just a kind of line with a high tension capability that is installed in the same path with correspondingly reduced sag. The transmission line(s) are then secured to the messenger cable every few feet so it assumes the shape of the messenger.

Keep in mind that the tension applied to the cable will also be applied to whatever anchor is supporting the end. This doesn't work if the tension pulls out the fasteners, or the wall they are attached to.

BUILDING ENTRANCE ARRANGEMENTS

One of the most common applications of transmission lines in Amateur Radio is the interconnection of antennas and radio equipment. By virtue of their operating characteristics, antennas tend to be outdoors and radio equipment indoors, requiring some way to get the line from one to the other. This can be a bit tricky in the case of housing — it happens that most houses are especially designed to not have holes in their walls for running cables.

Of course, there are already such penetrations provided for utility power, telephone service drop lines and coaxial cable for CATV or satellite TV service, as well as non-electrical facilities such as water, gas, sewage and heating flues. If you are the owner of the property, you have the option of observing the techniques used by those services and doing something similar for radio antenna connecting transmission lines, although some techniques may be much easier to accomplish before the dwelling is completed.

Drilling Holes in Walls

As the owner, or with the permission of an owner, you can usually drill holes in walls. I have done this many times and have learned a few things in the process.

Be Sure You Know What's Between Outside and Inside

There are few things worse for the amateur constructor than drilling through a wall and running through a water supply or water heating pipe. Of course these can be repaired, but usually the damage extends far and wide before you can get the water turned off. If you will need professional help to recover, a good operating rule is to not attempt something like that on a Sunday!

The best way to avoid such problems, be they caused by pipes or wiring, is to make sure you know what's there before you drill. One way that works for me is to make all such penetrations into unfinished space, such as the wall above a basement foundation. Coming in near ground level also has potential benefit in terms of your lightning protection plan, which we will discuss in the next section. My station is usually in the basement, so this is a natural; however, if it's not, it may be easier to make the basement entrance and then go between floors, since most floors either have no such obstacles, or they can be easily seen (more on this later).

If you don't have a basement, or if basement entry won't work for some other reason, you will need to come through in higher ground. Here it is very important to know what you are doing. If you are not familiar with construction techniques, find a house under construction that has similar characteristics. Look it over before they close up the walls. You will see some places that make sense to drill through. Note that (in my experience) every window and door is surrounded by a pair of 2 × 4 or other framing lumber. This means that a hole drilled next to a door or window, or beneath a window will be wood the whole way through — generally no pipes or wiring. In addition, you can probably come through a piece of molding, which is a good idea.

I like to drill so that there is wood on both sides, so the inside hole comes out in a wooden molding. This way the hole is solid and pulling cable won't rip up plaster or sheetrock. It is also very easy to patch after you are done, especially for painted woodwork — a few applications of spackle will usually make even the biggest hole disappear. If you want to be even more solid, drive a dowel in to just below the surface, to be a base for the spackle. Adding a bit of trim paint is much easier than repainting or papering the whole room. If next to a window, you often have the added benefit of having the cable entrance hidden by draperies.

Note that your usual twist drill will not be long enough for the task. Invest in a couple of 12-inch or longer drill bits in sizes that slowly come up to the cable size. An exterior wall will be the thickness of the framing studs, plus the sheetrock or paneling on the inside, plus the inside molding. On the outside you may have underlayment or sheathing and then multiple layers of shingles, for example. Another advantage of drilling through solid wood is that you can then just push the cable through the hole. If you are drilling through a hollow wall, you will have to snake your way through the insulation and then pull the wire through — not the end of the world, but certainly an extra hassle. Make sure you drill the hole at a slight angle downward going outward, so no water will run into the house through the hole or follow the cable in.

If You're Doing More Than One, Combine Them

Somehow it seems that most amateurs eventually have a need for more than one such penetration. Even if you start out with one, your interests and capabilities will likely grow as you get further into Amateur Radio. While one can just keep punching holes in different places, at some point some kind of entrance facility may make more sense. We'll discuss some in the context of lightning protection, but even independent of that, a single entrance facility may be a better idea. For example, interior and exterior electrical boxes on each side of the exterior sheathing and attached on the inside to a stud may make a professional-looking arrangement that is easy to add to just by popping off the covers on both sides. If you're not up to the task, having an hour of discussion time with a contractor or electrician may pay for itself in the long run.

Making a Window Entrance

If walls won't work, another possibility is a window entrance. While I have seen descriptions of ways to drill holes in glass, I've never attempted it. Instead, I've unglazed and removed the pane of glass, and replaced it with a piece of ⅛-inch polycarbonate sheet. The light transmission is just a bit less than through window glass, but not dramatically so. Still, I would avoid windows that highlight the appearance of a major room.

The polycarbonate is easy to work with using the usual tools, such as a reciprocating saw to cut it to size. Holes can be easily drilled, either for transmission line directly, or for PL-258 type or other feed-through connectors. If the holes are near the window edge, the cables may be able to be hidden by curtains or draperies as they go toward the floor. When it's time to move, recovery couldn't be simpler. Just remember where you stashed the pane of glass and swap it out with some glazing points and compound, and no one is the wiser.

Another possibility is to drill through the window sash itself. Carefully drill through a wooden sash in a spot that will be in the clear if the window is closed and that will not try to run through the glass. This can be accomplished with a short drill, is not likely to encounter plumbing or electrical runs and is easy to patch, if the surfaces are painted. While this is not quite in the same "no damage" category as the temporary board described below, it can be repaired to be almost invisible very quickly.

Temporary or "No Damage" Entrance Arrangements

If your circumstances are such that you can make no changes to the structure, or aren't even authorized to do any of this, there are still ways to bring antenna connections into the house. Some avoid the problem by using indoor or attic antennas. While that can keep you on the air, the chances are that you can have better results with an outdoor antenna. Here are a few possibilities.

Make the Connections Just When You Need Them

Most amateurs don't operate all the time, so they don't need an antenna connection all the time either. A transmission line connection can be just outside a window or door with its mate on the other side. When it's time to operate, just open the door, hook things up and leave the door open a crack while operating. Unless you are faced with heavy mosquito activity, this can work well. In fact some amateurs have the antenna, such as a short loaded vertical, just inside a patio door and move it all outside when it's time to operate — and when no one is looking.

Pinch Your Cable Under a Window Sash

Those using single-wire feed or window line can often put the line under a double-hung window and, if there's any gap or slop, just close the window on it, if it can still be latched. This is not ideal for a number of reasons, but is pretty easy to do. Perhaps a bit of "dressing" with a file can take the hard corner off the edge to make it less likely to damage the cable.

Add a Removable Panel Under a Window Sash

An improvement on the above that is almost as easy, is to get a small piece of trim board the width of your window sash and fit it under the lower sash, or above the upper sash so the window closes on it. Now you can notch (especially for cables with connectors) or drill as many holes as needed in your board for coax or other cables, or for feed-through connectors. If you don't want to build your own, MFJ offers a preassembled unit (see **Figure 9.8**) that can be cut to fit vertical or horizontal windows up to 48 inches wide and has feed-through arrangements for both coax and balanced lines. See **www.mfjenterprises.com**, and look for MFJ-4603.

There are two complications that need to be dealt with using this method. The first is that the security latch will no longer operate. This is pretty easy to fix. Either cut a piece of wood that can be jammed between the movable sash and frame, or install a small hardware store angle bracket so the window can't move. The screw hole can be easily spackled when it's time to move.

Figure 9.8 – The MFJ-4603 window entrance panel can be obtained with a number of feed-through connector configurations, or you can make your own from a piece of 1 × 2 lumber.

The other problem is that there will be an open passage between panes of the two sashes that will allow insects to come through. Mosquitoes are particularly prone to do so, since they are attracted by the focused emission of CO_2 from people breathing inside the house. This is the same problem that occurs with window air conditioners and yields to the same solution — just a piece of foam strip pushed into the space (get one from an air conditioner dealer, or in a pinch, it's a great use for rolled-up surplus T-shirts).

Drip Loops

With any cable entrance arrangement, it is critical that a drip loop be provided. This is easy to do; just make sure that the cable has a lower point before it enters the entrance arrangement. Otherwise, rainwater will follow the cable down and run into the wall or the feed-through mechanism. The drip loop allows a lower path from which the water will drip harmlessly onto the ground.

Extending the Route Between Floors or Rooms

If the entrance arrangement can't be on the same floor or in the same room as the station equipment, it is often fairly easy to continue on. If you can't make permanent holes, look for existing ones that you can follow. A building with central heating usually has ducts or pipes that run from room to room, or between floors. In my experience, there is often space to follow, especially with the thinner varieties of coax. If you can't push it through without losing it in the space between wall surfaces, try pushing a straightened coat hanger through from the other side. Then tape the coax to the hanger with electrical tape, making a point of the tape — along with tapering the coax end in extreme cases — so the coax won't be pushed off, and then pull it back.

If you can drill a path, you may be able to find a good path — again, be alert for the possibility of wiring or plumbing. An unfinished basement ceiling is the easiest to get through since it is only one surface thick (usually subfloor plus finished flooring, perhaps 2 inches total). My favorite technique is to go through under an interior wall with a molding. First checking to make sure you won't end up in a floor joist, start the hole through the molding at an angle to enter the basement slightly offset. This way results in an easy-to-patch hole in the molding, rather than having to deal with patching a hole in a finished floor.

Going between upper floors is a bit trickier to do without making a mess. In many cases, following existing penetrations may be the best way. If not, my next choice is to find closets one above another. A discreet hole between floors in a corner adjacent to the door, or even next to the door molding, is usually not too obvious, especially if the cable is then painted the same color as the wall surface. If you're not comfortable with any of this, a competent residential electrician can likely do it all for you, once he understands exactly what you want. If possible, do it in such a way that you can pull the cables out and replace them or add to them. Ask the electrician to leave a loose piece of cord along the path so you can add additional cables as your needs expand — and they will.

Bringing Balanced Transmission Lines In from the Cold

All the techniques that we discussed for coax can also be applied to balanced lines, with just a few caveats. Balanced lines need to avoid a number of situations that wouldn't bother coax. Conditions to watch for are runs that are within a few inches of metal ducting, pipes, or wiring. Also, lossy material that will absorb energy from the fields around the line must be avoided. Perhaps most critical is inadvertent coupling to and from signal lines, including telephone, network, or alarm systems.

If it is not possible to avoid such environments, it is possible to employ a section of dual coax balanced line, as discussed in Chapter 5 and illustrated in Figure 5.9. While this will be an unmatched section resulting in an impedance bump, that is not at all critical in the usual unmatched balanced line configuration, but should be considered if dealing with a matched condition, such as a 600 Ω feed to a rhombic antenna, or a 300 Ω feed to a folded dipole. The other consideration is that the loss will be that of unmatched coax, likely significant, so the length of this section should be as short as possible. For example, I used about 4 feet on one side of a rhombic feed as it went through a potentially lossy wall sill, just above the foundation line, without any problems.

GROUNDING AND LIGHTNING PROTECTION

While not a "transmission line" issue, per se, transmission lines that bring signals in from antennas can also bring in lightning energy if not properly protected. We will not go completely into lightning protection here, since that is arguably a subject for a book of its own; rather, we will briefly discuss ways to minimize what may be coming down the line as it comes inside.

A key to understanding lightning effects is to appreciate that with peak lightning strike current intensities from perhaps 20,000 A (50th percentile) to 200,000 A (90th percentile), it's not a question of where the current will go, but rather how much is left after you've gotten rid of as much as you can outside, before it heads toward the shack. This requires consideration of tower and antenna grounding with a low enough impedance to drain as much of the current as possible.

Still, it won't all go there — it will divide between paths to ground based inversely on impedance. Some will always go on the coax or other transmission line towards the station. It will show up in two forms; common mode current, essentially on the outside of the coax shield, and differential mode, between the inner conductor of the coax and the shield. In most environments, the common mode is a much larger fraction of the lightning current than the differential mode and can be reduced by low impedance grounding of the shields of coax as it comes into the building.

The Grounded Entrance Panel

The way to achieve this is to have the cable entrance at ground level and have as close to zero length conductivity as possible to a good ground structure at that point. This is not the usual solitary ground rod required by the National Electric Code (NEC), but a network of ground rods interconnected by buried, heavy, bare conductors that help distribute the current to the Earth's surface. The amount and extent needed for a desired grounding effectiveness will depend on the electrical characteristics of the soil at each location. The NEC, written into most local building codes, requires that all

Figure 9.9 – The grounded entrance panel at amateur station K8CH.

grounds be bonded together. Thus your radio ground system should be bonded to the ground used at the electric utility panel — preferably connected via a short heavy conductor (sized per the local building code) running outside the building.

The best way to provide maximum protection, in my opinion, is to have all cables enter through a common ground level entrance panel, as shown in **Figure 9.9**. By having the panel at ground level, it is possible to have the lowest impedance ground connection arrangement by having the ground system start right at the panel. Instead of using feed-through connectors, my ground level entrance panel (shown in Figure 8.3) has commercial lightning arresters that serve double duty and are bonded to an aluminum plate that is grounded to an extensive ground system tied to a rod at the window and extending outward to multiple rods through buried bare ground wires.

The window line entrance (not shown in Figure 8.3, but similar to the two feed-through insulators visible in Figure 9.9) goes to a balanced feed arrestor just inside the panel. It is also possible to use two coaxial lightning arrestors to protect the equipment tied to the balanced feed line, with one conductor on each arrestor. Just make sure you calculate how high the operating voltage will be, and select appropriate arrestors so that they won't fire during normal transmission.

At my station, I have push-on type UHF coax connectors on my coax cables going to the inside connection of the arrestors (available from **www.americanradiosupply.com** and others) and double banana type going to the balanced arrestor. I routinely keep them disconnected when not actually on the air. In addition, I keep my ARRL All Risk equipment insurance policy up to date (see **arrlinsurance.com**).

Above-Ground Entrance Arrangements

A station located on an upper story without ground-level access has a different set of challenges. While it may be tempting to reduce transmission line length by making runs directly to the station, arguably it is the worst case from a lightning protection perspective, since all lightning currents on transmission lines are forced to go through the station. A better arrangement is to run the cables to ground level with an appropriate grounding and surge protection interface at that point, as shown in **Figure 9.10**. It's not bad to have additional arrestors at the station end, but the ground-level ones are the most important, in my view.

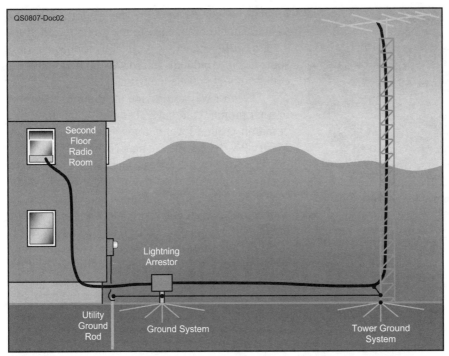

Figure 9.10 – Suggested arrangement for lightning protection grounding of an upper-story cable entrance. By running the cables first to grounding and lightning protection at ground level, the lowest impedance ground connection is provided.

TRANSMISSION LINE CARE AND MAINTENANCE

Fortunately, transmission lines tend to be rather low maintenance, however, that's not quite the same as "no maintenance." Transmission lines can last for many years, but none last forever, with typical expected useful lives of 10 to 20 years. That, of course, assumes proper installation and care.

Premature Line Degradation

Environmental factors are the major cause of early line degradation. Probably the most significant are the result of water penetration that occurs due to insufficient sealing of ends, or through jacket penetration resulting from abrasion. The insidious aspect of this is that such gradual degradation is not easily noticed through normal operation until it goes quite far. While we will pay lots of money for an additional 2 or 3 dB of antenna gain, we are not too likely to notice 2 or 3 dB of additional transmission line loss. In fact, the additional line loss results in lower SWR at the station end, along with easier tune up and wider bandwidth — each of which could be, but probably aren't, noticed due to the gradual change.

Line Inspection

A regular program of physical inspection of transmission lines is most appropriate. Areas that are becoming abraded, show signs of insulation cracking, or from which waterproofing material is becoming unsealed can be readily addressed before they cause serious problems. A good application of electrical tape can provide additional coverage, as well as additional layers in areas subject to abrasion.

Be aware of areas that are bent more tightly than the allowed bending radius shown in Table 9.1. Foam dielectric cable, while lower in loss than cables with solid polyethylene dielectric, is subject to migration of the center conductor and possible shorting, if bent too much.

Another potential problem area is any solder connections that are exposed to the elements. Solder oxidizes in nature and can erode over time, eventually making for intermittent or failed connections.

Routine Measurement

It is a good idea to take and record a complete SWR run across frequencies of each antenna and transmission line system at installation. If nothing degrades, the values should stay unchanged over time. As a general rule, if the SWR goes up, look to the antenna for changes. If the SWR goes down, check for additional transmission line loss, since antennas rarely retune themselves, left to their own devices.

Checking Line Operation

Perhaps the best way to evaluate a line's operational characteristics is to measure the loss and SWR before installation, and then again during regular inspection periods. Just put a signal into a dummy load and wattmeter at the radio end, without the line and repeat the measurement at the far end. Any reduction in power delivered should be noted, and the loss calculated and recorded. Additional loss over time is an indication of degradation.

Checking Line Loss from One End

While measuring loss as described above is the most accurate method, sometimes it is not feasible to get equipment to the far end. In that case we can get a good measure of line loss from one end by measuring the SWR at the frequency of interest with the coax either open or shorted. If we had ideal lossless coax, our measured SWR would be infinite, since all the power would be reflected and would return to the source. With losses

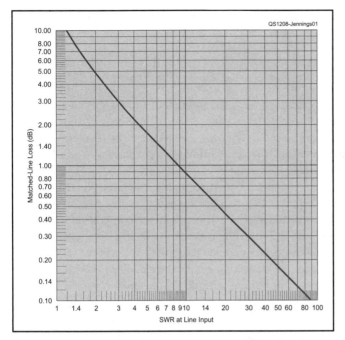

Figure 9.11 – Graph of the apparent SWR versus losses in a coax measured terminated in an open or short. Find the SWR value you measured on the horizontal axis, then go straight up and read the one way matched loss in dB on the vertical axis where the SWR reading crosses the diagonal line. Note that if the line is not matched, the loss can increase significantly.

in the coax, something else happens. Whatever signal power we put into one end will arrive at the other end with the same power, less the amount of line loss. Because the short or open at the far end of the coax provides a total reflection, the signal then reverses its direction and returns to the source end. On the return trip, the power level at the far end of the coax will be again reduced by the losses in the coax.

The reduced reflected power will show up in real coax as an SWR that is less than infinite. The higher the loss in the coax, the lower the measured SWR as read at the test end. Carried to extremes, a very high loss in a length of coax (such as the losses at UHF for a 500 foot long piece of RG-58) will result in an SWR of nearly 1:1 with the far end of the coax open or shorted.

Figure 9.11 shows a graph from the 16th edition of *The ARRL Antenna Book* of the SWR versus matched loss in a coax measured with a short or open at the far end. The section also provides an equation to determine the one-way matched loss from the measured SWR: L_M is the one-way matched line loss in decibels at the measurement frequency.

$$L_M = 10 \times \log_{10} \{(SWR + 1)/(SWR - 1)\}$$

In other words, if you measured an SWR of 3:1 with a length of open or shorted coax, either Figure 9.11 or the above equation indicates that the cable loss will be 3 dB at the frequency of interest.

To make the measurement, first make sure the antenna end of the coax is not connected to anything, and that any shorts you placed there during the ohmmeter tests have been removed. Follow the instructions in the manual for the antenna analyzer to make the SWR measurement.

As an example, let's say we measured the SWR of a 50 foot length of

RG-213 used at a dual-band VHF/UHF repeater site. At 147 MHz the open circuit SWR was 6.6:1. This indicates a loss of about 1.3 dB, compared to 1.4 dB predicted by *TLW* for 50 feet of RG-213 (Belden 8267).[2] At 445 MHz, the measured SWR was 1.4, indicating a loss of 7.78 dB, much worse than the 2.6 dB predicted by *TLW*. Thus this coax, while useful on 2 meters, will not be very good on 70 cm and is probably in the process of degrading.

What if the Antenna is Up in the Air?

The tests described previously are great if you have access to both ends of the coax, but sometimes you would like to find out about that coax without going up the tower. While you don't have control over the far end, you likely know what's there, if it's your antenna. For example, I have a 2 meter Yagi with a T matching section. While that provides a nice 50 Ω termination at 2 meters, it should look like a short at dc, and an ohmmeter continuity test should show a very low resistance, almost as low as if it were a short. A split feed Yagi or dipole should look like an open at dc.

The SWR at 2 meters won't tell us much about loss, in fact the lossier it is, the better the SWR will look. On the other hand, it will look a lot like a short at $\frac{1}{10}$ the frequency or at 20 meters. A measurement there should give us a good, but not precise, idea of the 14 MHz loss. If the cable is good there, chances are it is also good at 2 meters. Other types of antenna connections may be trickier to make use of, but if you measure the SWR on other frequencies when the cable is new, and store the data in your archives, any change in later years may mean either the antenna or transmission line has undergone a change that merits investigation.

Another trick is to sweep your analyzer frequency looking for the worst SWR. That probably indicates either a very high or very low impedance termination and the loss at that frequency will be no better than that indicated by the SWR.

A Cautionary Note — Note that all of the loss results based on measured SWR are no better than the data provided by the SWR analyzer. *The ARRL Antenna Book* notes: "The instruments available to most amateurs lose accuracy at SWR values greater than about 5:1, so this method is useful principally as a go/no go check on lines that are fairly long." That really is the only question you need answer — do we want to keep this coax, or is it time for a replacement? These tests should give you enough information to make those decisions.

Notes
[1] L Nelson, AB0DZ, "Hints & Kinks — Anchoring Coaxial Feed Line," *QST*, Aug 2009, p 60.
[2] *The ARRL Antenna Book,* 16th Edition.
[3] *TLW, Transmission Line Program for Windows* software is provided on a CD with *The ARRL Antenna Book*.

Index

A
Amateur Radio station 1-1
   Transmission line applications 1-2
Attenuation of matched transmission lines .... 3–10 ff
8–2
Attenuation of mismatched transmission lines .... 3–14 ff

B
Balanced line interconnections
   Double banana plug and jack 6–14
   TV twinlead connectors 6–14
Balanced microstrip line 2–6
Balanced to unbalanced (balun) transformer 8–3
Balanced transmission line
   Attenuation 3–12
   Balanced microstrip line 2–6
   Electrical characteristics of balanced line 5–8
   Fields between conductors of balanced line 5–7
   Matched line loss of balanced line .... 5–8 ff
   Open-wire line 2–3
   Parallel coax line 2–6
   Power handling capability of balanced line 5–9
   Shielded twisted pair 2–5
   The benefits of balanced line 5–6
   The downsides of balanced line 5–7
   Twinlead 2–4
   Twisted pair 2–5
   Types of balanced line 5–3
   Window line 2–5
Building entrance arrangements
   Drilling holes in walls 9–5
   Drip loops 9–8
   Making a window entrance .... 9–6 ff
   Temporary arrangements 9–7

C

Center-fed dipole .... 1–2 ff
Characteristic impedance .... 3–3, 3–10 ff, 8–2
Coaxial transmission line
    Additional loss due to mismatch .... 3–14 ff
    Attenuation .... 3–10 ff
    Characteristic impedance .... 3–3 ff, 3–10 ff
    Configuration .... 2–1 ff
    Direct burial 8–6
    Equivalent circuit .... 3–1 ff
    Fields within coax .... 3–6 ff
    Key cable parameters .... 3–9 ff
    Matched loss .... 3–13 ff
    Shielding limitations .... 3–7 ff
    Types of coax cable .... 3–8 ff
Concelman, Carl 6–8
Connectors for transmission line
    Balanced line interconnections .... 6–13 ff
    Between series adapters 6–12
    BNC coaxial connectors .... 6–8 ff
    F-type connectors 6–10
    General Radio connectors 6–12
    Limitations of Type N coaxial connectors .... 6–5 ff
    Limitations of UHF series coaxial connectors 6–4
    Motorola connectors .... 6–11 ff
    RCA connectors 6–9
    SMA connectors 6–10
    TNC coaxial connectors 6–11
    Type C coaxial connectors 6–11
    Type HN coaxial connectors 6–12
    Type N coaxial connectors 6–5, 7–9
    UHF series coaxial connectors 6–2

D

Double banana plug and jack 6–14
Drilling holes in walls 9–5
Duffy, Tim, K3LR 7–5

F

Fields between conductors of balanced line 5-7
Fields within coax .... 3–6 ff

G
Grounding and lightning protection 9–10
Guided wave structures
    Single wire transmission line 2–7
    Waveguide 2–7

I
Installing coax BNC connectors 7–14
    Installing BNC connectors 7–14
    Installing RCA (phono) connectors 7–16
    Installing Type F connectors 7–14
    Installing Type N connectors .... 7–9 ff
    Installing UHF series connectors .... 7–2 ff
    Standard or crimp .... 7–1 ff
Installing Type N connectors
    UHF-like method 7–13

L
Lightning protection 9–10

M
Microstrip transmission line .... 2–2 ff

N
Neill, Paul 6–5, 6–8

O
Outdoor use of transmission lines .... 8–4 ff

P
PL-259 .... 7–3 ff
Power rating of transmission lines 8–2

R
RF energy 1–1
RG-11 coax 7–3
RG-213 coax 7–3
RG-214 coax 7–3
RG-58 coax 7–7
RG-59 coax 7–7
RG-8 coax 7–3
RG-8X coax 7–7

S

Sag in line runs 9–4
Satellite TV
    System digram 1–4
SMA coaxial connectors 6–10
SO-239 UHF socket 7–8
    Shielding hood 7–9
Stripline transmission line 4–3

T

TNC coaxial connectors 6–10
Transmission line care and maintenance
    Checking line operation and loss .... 9–13 ff
    Line degradation 9–12
    Line inspection 9–13
    Routine measurement 9–13
Transmission line in RF environment 8–6
Twinlead .... 2–4 ff
Type N connector
    Version 7–10

U

UG-175/U .... UG-176, 7–7 ff
UHF series coaxial connectors .... 6–2 ff
Unbalanced transmission line
    Coaxial cable .... 2–1 ff
    Coplaner strip transmission line 4–4
    Microstrip transmission line .... 2–2 ff, 4–1
    Single wire isolated from ground 4–5
    Single wire line 2–3
    Single wire over a ground plane 4–4
    Stripline transmission line 4–3
UT-8000 coax connector tool 7–3

W

Waveguide 2–7
Windom antenna 4–6
Window line 2–5

FEEDBACK

Please use this form to give us your comments on this book and what you'd like to see in future editions, or e-mail us at **pubsfdbk@arrl.org** (publications feedback). If you use e-mail, please include your name, call, e-mail address and the book title, edition, and printing in the body of your message. Also indicate whether or not you are an ARRL member.

Where did you purchase this book?
☐ From ARRL directly   ☐ From an ARRL dealer

Is there a dealer who carries ARRL publications within:
☐ 5 miles   ☐ 15 miles   ☐ 30 miles of your location?   ☐ Not sure.

License class:

☐ Novice

☐ Technician

☐ Technician with code

☐ General

☐ Advanced

☐ Amateur Extra

Name _____	ARRL member? ☐ Yes ☐ No
	Call Sign _____
Daytime Phone ( ) _____	Age _____
Address _____	
City, State/Province, ZIP/Postal Code _____	
If licensed, how long? _____	E-mail _____
Other hobbies _____	
Occupation _____	

For ARRL use only	CFTL
Edition	1 2 3 4 5 6 7 8 9 10 11 12
Printing	1 2 3 4 5 6 7 8 9 10 11 12

From _____
_____
_____

Please affix postage. Post Office will not deliver without postage.

EDITOR, THE CARE AND FEEDING OF TRANSMISSION LINES
ARRL—THE NATIONAL ASSOCIATION FOR AMATEUR RADIO
225 MAIN STREET
NEWINGTON CT 06111-1494

— — — — — — — — — — — — — please fold and tape — — — — — — — — — — — — —